Wind Power and Public Engagement

Adopting an interdisciplinary social science approach, this book examines community reactions to wind farms to form a new understanding of what facilitates social acceptance.

Based on empirical research, *Wind Power and Public Engagement* investigates opposition to wind energy and considers the advantages as well as the limits of the co-operative model of wind farm community ownership. Giuseppe Pellegrini-Masini compares the role of co-operative schemes with community benefits schemes in increasing acceptability, and also sheds light on the impact of social factors including pro-environmental attitudes, perceived benefits and costs, place attachment, trust, as well as individuals' resources such as information and income. Five research cases are investigated in England and Scotland, including the first local, community-owned wind farm co-operative in the UK. Critically reviewing existing social research theories, the book offers a new viewpoint, integrating rational choice and environmental attitudinal theories, from which to assess and understand the social acceptability of wind energy. It also highlights new opportunities for raising consensus in communities around locally proposed wind farms.

The book will be of great interest to students and scholars of renewable energy, energy policy, environmental sociology, environmental psychology, environmental planning and sustainability in general, as well as policymakers.

Giuseppe Pellegrini-Masini is a Postdoctoral Fellow in the Department of Psychology at the Norwegian University of Science and Technology (NTNU), Norway, and a Research Affiliate in the Research Area in Public Ethics at the Sant'Anna School of Advanced Studies, Italy. He holds a PhD from the School of Energy, Geoscience, Infrastructure and Society at Heriot-Watt University, Scotland.

Routledge Studies in Energy Policy

For further details please visit the series page on the Routledge website: www.routledge.com/books/series/RSIEP/

Wind Power and Public Engagement

Co-operatives and Community Ownership

Giuseppe Pellegrini-Masini

 Routledge
Taylor & Francis Group

LONDON AND NEW YORK

 earthscan
from Routledge

First published 2020
by Routledge
2 Park Square, Milton Park, Abingdon, Oxon OX14 4RN

and by Routledge
605 Third Avenue, New York, NY 10017

First issued in paperback 2022

Routledge is an imprint of the Taylor & Francis Group, an informa business

Publisher's Note
The publisher has gone to great lengths to ensure the quality of this reprint but points out that some imperfections in the original copies may be apparent.

British Library Cataloguing-in-Publication Data
A catalogue record for this book is available from the British Library

Library of Congress Cataloging-in-Publication Data
A catalog record has been requested for this book

ISBN: 978-0-367-50304-8 (pbk)
ISBN: 978-1-138-58910-0 (hbk)
ISBN: 978-0-429-49189-4 (ebk)

DOI: 10.4324/9780429491894

Typeset in Times New Roman
by Wearset Ltd, Boldon, Tyne and Wear

Contents

Figures

Tables

Acknowledgements

I wish, above all, to thank Professor Phillip Frank Gower Banfill of Heriot-Watt University, without whose enduring support this work would not have been completed. He demonstrated to me the most important skills that a supervisor should possess: respect, trust and empathy. I also thank my family, my friends and my colleagues for the support and advice they gave me at various times during the years when I was working on this project. Finally, I wish to thank Professor Christian A. Klöckner in the Department of Psychology at Norwegian University of Science and Technology (NTNU), who saw the merits of this work and lent his time to write the Foreword.

This research was carried out thanks to the support of the following projects: EPSRC SUPERGEN "Future Network Technologies", EPSRC TARBASE "Technology Assessment for Radically Improving the Built Asset base", and Horizon 2020 SMARTEES "Social Innovation Modelling Approaches to Realizing Transition to Energy Efficiency and Sustainability", grant agreement No 763912.

Foreword

When the United Nations Sustainable Development Goals were accepted by the General Assembly of the UN in 2014 the world community committed to a list of seventeen common goals, to be reached by 2030. This included goals such as the ending of poverty, affordable and clean energy, reducing inequality, gender equality, sustainable cities and communities, responsible consumption and production and climate action. To reach these goals, a change in the way energy is produced and consumed in both established and upcoming economies is of crucial importance; the energy system is the backbone of our societies. The centralised energy systems that drove the industrial revolution in Western and Eastern economies in the twentieth century have clearly reached their limits. New organisational structures need to be found to tackle the challenges of the future, including a high degree of regenerative (and often unstable) energy sources, decentralisation of energy production, a much more active role for energy consumers and a complete shift in the roles of actors in the energy system.

These substantial changes in established systems are at the very basis of what actors expect from themselves and each other in these systems. Thus, the question of how to master these transitions is not so much technological or economic, but rather a question of understanding (local) cultures, practices, values and social rules. The psychology of decision makers at all levels moves into the focus of analysis, so we are able to understand potential facilitators or barriers in these transition processes.

With only ten years to go before we must reach the Sustainable Development Goals, and also only one decade left to limit global climate change to "substantially below 2 degrees" by implementing massive cuts in CO_2 emissions, speeding up the energy transition is essential. This book highlights the new complexity of energy decisions through a comprehensive analysis of community wind farm co-operatives in the United Kingdom. It goes way beyond analysing ecological, economic and technical challenges and focuses on what co-operative models of energy production contribute to social sustainability, energy justice or democratisation of the energy system. This has clear impacts on the acceptability of energy projects that can have an invasive character in local communities, such as onshore wind parks.

The empirical work behind this book shows that community-driven energy production moves the focus from purely economic evaluations to include a more diverse set of criteria, balancing economy with negative local effects on nature or health with benefits for the global climate and considerations for alleviating fuel dependency. Those engaged in creating local co-operatives obviously consider a large set of local to global costs and benefits, which makes their judgments potentially more acceptable for local communities than decisions made by external commercial actors. A key benefit of co-operative forms of energy production for local communities seems to be that it offers community members the option of influencing operations rather than just benefitting (or suffering) from such implementation. Co-operative production thus gives community members stronger agency.

Although it appears that co-operative energy production is a viable way to greater engagement of local communities in the energy transition, it is not a silver bullet that will solve all the issues that arise with the necessary restructuration. From a bigger picture perspective, it is also interesting to interpret the results of this work concerning energy justice questions, especially distributional and procedural justice. Are co-operative energy producers per se more trustworthy with respect to just distribution of energy or revenues? Do they guarantee more just procedures in implementing projects? The Conclusions of this book analyse such questions from a policy-making perspective. *Wind Power and Public Engagement* thus makes a valuable contribution to some of the biggest challenges our societies have to face in the coming ten years – enabling communities to make fast and substantial changes.

Christian A. Klöckner
Professor in Social Psychology and Quantitative Methods
Norwegian University of Science and Technology

Introduction

While the transition to a new energy system based on the wide deployment of renewables has become an undisputed necessity for our societies, many questions remain in terms of how energy should be delivered. Will nations and regions adopt policies that seek to maximise energy justice (Jenkins *et al.*, 2017) and energy equality (Pellegrini-Masini, 2019) through diffused ownership of renewable energies? Or will they reproduce the concentrated ownership of the current energy production system?

Technologies such as solar photovoltaic, whose middle-class affordability has already been demonstrated, will most likely play a role in the energy transition towards an energy system centred on prosumerism. Yet this might not be the case for large solar plants, wind turbines or hydroelectric systems, all of which present substantial upfront costs that are a barrier to citizen ownership.

Still, the possibility for citizens to be part of the energy transition, not only as consumers but also as owners, is a significant step towards a more just energy system and one that reduces energy inequalities.

We know that energy consumption is largely unequal, not only across countries but also within them (Pachauri and Spreng, 2012). This trend reflects differences in income distribution, but it also reveals a vast problem with environmental equality and justice.

As the UN Human Rights Council (2019) has compellingly stated:

> Perversely, the richest, who have the greatest capacity to adapt and are responsible for and have benefitted from the vast majority of greenhouse gas emissions, will be the best placed to cope with climate change, while the poorest, who have contributed the least to emissions and have the least capacity to react, will be the most harmed. The poorest half of the world's population—3.5 billion people—is responsible for just 10 percent of carbon emissions, while the richest 10 percent are responsible for a full half. A person in the wealthiest 1 percent uses 175 times more carbon than one in the bottom 10 percent.
>
> (p. 6)

This situation will create a significant and challenging policy problem for years to come. Perpetuating this state of affairs would be at odds with the core values of Western democracies and the procedural or formal equality principle that defines democratic systems and processes. Further, it would run counter to the substantive or distributional equality principle that is to some extent guaranteed by many democracies, whose constitutions provide for the right of citizens to satisfy their basic needs.

Recent conceptualisations of energy equality (Pellegrini-Masini, 2019) have argued for the need to establish greater equality of opportunity in using energy services and consuming energy or embodied energy. The desirability of greater "equality of opportunity" can be defended on both moral (Sen, 1980) and empirical (Wilkinson and Pickett, 2009) grounds. Egalitarian societies earn positive high rankings regarding many different quality-of-life indicators, ranging from health to crime to environmental care (Wilkinson and Pickett, 2009) and high-income countries appear to produce fewer emissions per capita (Jorgenson *et al.*, 2015; Jorgenson *et al.*, 2016; Knight *et al.*, 2017). Despite some concern regarding the economic efficiency of egalitarian societies (Okun, 2015), recent research shows that high inequality negatively affects economic growth and that redistributive policies do not hinder growth (Cingano, 2014).

Arguably, the energy transition should not only deliver environmental sustainability but should also be economically and socially sustainable, thereby addressing energy inequality and delivering energy justice more broadly.

The goals of equitable, sustainable energy have already been put forward by European states, as the EU Commission (2019, p. 6) recently confirmed: "The clean energy transition must be fair and socially acceptable to all".

Community energy initiatives, particularly community wind farm co-operatives, are a way to contribute to a new economy of energy production in which pro-sumers (consumers who participate in the production process) are the norm rather than the exception. However, co-operatives are not devoid of challenges and can face similar obstacles to those experienced by other types of community energy initiatives. Specifically, community energy schemes risk creating a divide between individuals who have the financial means and the ability to join them and those who do not (Jenkins, 2019).

This book adopts an interdisciplinary social science approach to examine the factors that influence the acceptability of wind farms and wind farm co-operatives. Specifically, it addresses the following questions: What are the variables that influence participation in wind farm-co-operatives? Can wind farm co-operatives facilitate the acceptance of wind farm development at the pre-construction stage? How do co-operatives compare with other community wind farm schemes? What can be learned about the social acceptability of wind farms and renewable energy infrastructure more generally from previous research on the acceptability of wind farms and wind farm co-operatives? And finally, how might policy address resistance to renewable energy infrastructure?

Chapter 1 reviews the literature on citizen participation, activism, and pro-environmental behaviours that theoretically frame the extensive range of variables that affect participation in wind farm co-operatives. In fact, participation in a wind farm co-operative can be both a pro-environmental behaviour and a type of environmental activism.

Chapter 2 presents the specific variables that influence citizens' participation in wind farm co-operatives. Variables regarding types of perceived costs and benefits associated with wind farm siting, along with contextual and personal resource variables, are discussed. Further, the chapter examines co-operative schemes within the context of community energy schemes and their implications for wind farm acceptability. Finally, the theoretical framework that informed the empirical research presented in the book is outlined.

Chapter 3 presents the multimethod research design, based on a triangulation approach, along with details about the cases surveyed and the specific methods used.

Chapter 4 analyses interviews conducted with stakeholders of Westmill Wind Farm in Oxfordshire, Britain's first community-owned co-operative wind farm. The Westmill case provides compelling evidence on the strengths and the limits of the co-operative model, and reveals the profound impact that variables such as trust can have in shaping local debates and influencing levels of acceptability.

Many of the variables discussed in the analysis of the qualitative study also appear in Chapter 5, which presents the results of a multi-site survey conducted in four Scottish cases. The chapter provides a quantitative analysis of the wide range of variables, detailing how they correlate with the acceptability of proposed wind farms and their relative importance in influencing acceptability. It also examines additional variables regarding respondents' opinions about the co-operative model. The large number of variables surveyed (57) and their nature make this research distinctive, although they restricted the scope of the analysis to a limited range of statistical tests.

Chapter 6 discusses the survey results, examining the different groups of variables surveyed and, within each group, the specific variables that chiefly influenced opinions about the wind farm and the co-operative model. While some earlier assumptions about the role of pro-environmental attitudes are challenged, the chapter underlines the importance of perceived costs and benefits and, above all, "trust" as key determinants of the acceptability of wind farms.

While much can be said in support of co-operative wind farms and, to some extent, against them, this research shows that respondents were predominantly favourable towards wind farm co-operatives and to being offered the opportunity to participate in such a scheme, along with the provision of community benefits.

Certainly, community benefits are not a secondary issue. Instead of dismissing them as an unworthy policy provision of dubious morality, which could "crowd-out" environmental morale, it is time to advocate for a complementary approach (Frey, 1999) that strengthens pro-environmental attitudes and extrinsic motivations towards the goal of making onshore wind farms more acceptable. Such a pragmatic approach would be anything but immoral, not only because it would favour an overdue energy transition, but also because it could address issues of energy justice at various levels, including that of citizen compensation for facility siting (Corvino *et al.*, 2019).

References

Cingano, F. 2014. *Trends in income inequality and its impact on economic growth.* OECD Social, Employment and Migration Working Papers. OECD Publishing, Paris.

Corvino, F., Pirni, A., Pellegrini-Masini, G. and Maran, S. 2019. Compensation for energy infrastructures: Can a capability approach be more equitable? Unpublished manuscript, Dirpolis Institute, Sant'Anna School of Advanced Studies, Pisa, Italy.

European Commission. 2019. *Clean energy for all Europeans.* Directorate-General for Energy, Luxembourg. Available at: https://publications.europa.eu/en/publication-detail/-/publication/b4e46873-7528-11e9-9f05-01aa75ed71a1/language-en.

Frey, B. S. 1999. Morality and rationality in environmental policy. *Journal of Consumer Policy*, 22, 395–417.

Jenkins, K. E. H. 2019. Energy justice, energy democracy, and sustainability: Normative approaches to the consumer ownership of renewables. In: Lowitzsch, J. (ed.) *Energy transition: Financing community co-ownership in renewables.* Cham, Switzerland: Palgrave Macmillan.

Jenkins, K., McCauley, D. and Forman, A. 2017. Energy justice: A policy approach. *Energy Policy*, 105, 631–634.

Jorgenson, A. K., Schor, J. B., Huang, X. and Fitzgerald, J. 2015. Income inequality and residential carbon emissions in the United States: A preliminary analysis. *Human Ecology Review*, 22, 93–106.

Jorgenson, A. K., Schor, J. B., Knight, K. W. and Huang, X. 2016. Domestic inequality and carbon emissions in comparative perspective. *Sociological Forum*, 31, 770–786.

Knight, K. W., Schor, J. B. and Jorgenson, A. K. 2017. Wealth inequality and carbon emissions in high-income countries. *Social Currents*, 4, 403–412.

Okun, A. M. 2015. *Equality and efficiency: The big tradeoff.* Washington, DC: Brookings Institution Press.

Pachauri, S. and Spreng, D. 2012. Towards an integrative framework for energy transitions of households in developing countries. In: Spreng, D., Flueler, T., Goldblatt, D. L. and Minsch, J. (eds.) *Tackling long-term global energy problems.* Dordrecht, Netherlands: Springer.

Pellegrini-Masini, G. 2019. Energy equality and energy sufficiency: New policy principles to accelerate the energy transition. European Council for an Energy Efficient Economy 2019 Summer Study, "Energy efficiency first, but what next?", 3–8 June, Belambra Presqu'ile de Glens, France, 143–148.

Sen, A. 1980. Equality of what? In: McMurrin, S. M. (ed.) *The Tanner lectures on human values.* Cambridge: Cambridge University Press.

UN Human Rights Council. 2019. *Climate change and poverty: Report of the special rapporteur on extreme poverty and human rights.* United Nations, Geneva. Available at: https://srpovertyorg.files.wordpress.com/2019/06/unsr-poverty-climate-change-a_hrc_41_39.pdf [Accessed 25 June 2019].

Wilkinson, R. G. and Pickett, K. 2009. *The spirit level: Why greater equality makes societies stronger.* New York: Bloomsbury Press.

1 Citizens and renewable energy
Determinants of civic engagement

Community social acceptability of wind energy

One of the primary aims of this work is to discuss the social acceptability of wind farms in a community setting. Social acceptability will lead to acceptance and acceptance to support, resulting in civic engagement.

The use of the term *acceptability* rather than *acceptance* or *support* is deliberate, essentially motivated by the literal meanings of the words. Acceptability means "the quality of being satisfactory and able to be agreed to or approved of" (Batel *et al.*, 2015), while acceptance means "general agreement that something is satisfactory or right, or that someone should be included in a group" (Cambridge Dictionary, n.d.). This book will focus on the qualities of proposed wind farms that made them acceptable and, specifically, on community co-operative schemes. In his discussion of wind technology, Szarka (2007, p. 17) makes the same semantic choice:

> The question of acceptability is not the same as the question of acceptance. In the context of this book, the issue of acceptability is not addressed to create the presumption that wind power per se is unacceptable. Rather, the question concerns the criteria and conditions under which a social, economic or institutional actor decides to accept or reject an idea, vision, proposal or practice.

Wüstenhagen *et al.* (2007) distinguish between three dimensions of social acceptance: community, socio-political and market acceptance, which can be translated to the concept of acceptability. This work will focus on the dimension of community acceptability. Therefore, social acceptability will specifically refer to community acceptability, unless otherwise indicated. The community dimension might appear self-explanatory; however, for purposes of clarity, it is worth noting that the term community will refer to a set of local stakeholders, including—first and foremost—local residents, but also including local authorities and local organisations (Wüstenhagen *et al.*, 2007).

Policy research context: energy policy in the United Kingdom

The 2003 UK Energy White Paper (DTI and DEFRA, 2003) stated that renewable energy and distributed generation expansion, as well as energy efficiency, were the best methods for achieving substantive CO_2 reduction. Later, the UK government (HM Government, 2009) reiterated its commitment to the wide deployment of wind-generated energy to reduce carbon emissions and achieve the legally required target of sourcing 15 per cent of the UK's energy from renewables (DECC, 2011). In 2019, generic support for renewables was reiterated for the whole of the UK, while acknowledging the national targets of Scotland (50 per cent of all heat, transport and electricity generated by renewables) and those of Wales (70 per cent of electricity from renewables including 1 gigawatt of installed community-owned renewable energy by 2030) (Department for Business, Energy & Industrial Strategy, 2019)

The 2003 Energy White Paper noted that "[i]ncreasing the deployment of renewables will depend on people supporting local projects." Moreover, it insisted that there is "a clear benefit in local communities becoming producers, as well as consumers, of energy, establishing and benefiting from the local ownership of some forms of generation" (DTI and DEFRA, 2003, pp. 51–52). Therefore, along with the Renewables Obligation that requires electricity suppliers to deliver a specified amount of electricity originating from eligible renewable sources, specific policy instruments were implemented that favoured community energy initiatives, such as the Countryside Agency's Community Renewables Initiative, launched in 2002.

The UK government has renewed this commitment with the "Community Energy Strategy" (DECC, 2014, 2015), which states:

> Our ambition is that every community that wants to form an energy group or take forward an energy project should be able to do so, regardless of background or location. We will back those who choose to pursue community energy, working to dismantle barriers and unlock the potential of the sector.
>
> (DECC, 2014, p. 7)

In this context, Scotland has used its devolved powers to develop an even more ambitious energy policy than the one drafted for the whole of the UK. The Scottish government has recognised the need to involve local communities by promoting participation in commercial schemes and community ownership (Scottish Government, 2011). Further, it has set a target of 500 MW of installed capacity for community and locally owned renewables by 2020 and has encouraged developers to include an element of shared ownership for every renewable energy project development above 50 kilowatts (Local Energy Scotland, 2015).

Renewable energy and the future electricity system

Renewable technologies are perceived as an important method of delivering an energy future with reduced greenhouse gas (GHG) emissions and increased environmental protections (DTI and DEFRA, 2003; HM Government, 2009; Department for Business, Energy & Industrial Strategy, 2019). Such a view of an energy future is shared by many authors, including Lovins (1977),

Patterson (1999) and Rifkin (2002). All of these scholars foresee a future involving a large deployment of distributed generation, with the transition from a centralised and hierarchical electricity system towards a decentralised electricity system that will involve citizens as producers (Rifkin, 2002). This transition will contribute to an increasingly democratic system of generating electricity as well as a more democratic society in general. Rifkin (2002) considers the possibility that, in the future, distributed generators owned by families, neighbourhoods and communities will be united in DGAs (distributed generation associations), using the cooperative model developed in the nineteenth century to protect consumers' rights and building a retail structure able to deliver goods to the market at a lower, more affordable price.

Citizenship policy

Former UK Home Secretary David Blunkett (2003) made clear the importance of fostering active citizenship in local communities in order to successfully deliver public goods such as environmental quality, security and youth care. The UK government therefore implemented this policy position through instruments such as Community Interest Companies (CICs) that were meant to be a new form of social enterprise and launched in July 2005 (The CIC Project Team, 2004). A similar ethos of active citizenship was also promoted by succeeding conservative governments, albeit rebranded as the "big society" (Dinham, 2010). Local renewable energy co-operatives can be viewed as a form of this active citizenship ethos because they seek to enhance global environmental quality, contribute to local wealth and strengthen community bonds, thereby increasing local social capital (Lipp and McMurtry, 2015).

Citizenship

The traditional meaning of *citizenship* is membership in a political community that entitles a subject to the exercise of rights and obliges him/her to fulfil some duties (Reeve, cited in McLean and McMillan, 2009).

Prior *et al.* (1995, p. 5) describe citizenship thus: "In our view one of the confusions in current discussions of citizenship arises from a failure to recognise that there are two distinct dimensions to the concept", that is, "... citizenship as a status which people possess ... [and] citizenship as a practice which people engage in." Therefore, "Citizenship is thus a concept both of being and doing".

Therefore, citizenship can be considered not only as a status (the traditional meaning in political thought), but also as a practice, originating from the entitlements that the status provides. Consequently, it is possible to include every aspect of a subject's public life in the definition of citizenship.

According to Faulks (2000), the definition of citizenship is constantly evolving. In this respect, it is worth mentioning that two main philosophical views have influenced, and still influence, this idea: libertarianism and social democratic political thought, in accordance with Prior *et al.* (1995).

Libertarianism views citizenship as reduced to a limited public role in society; in fact, subjective action within a society is viewed as predominantly expressed through participation in free market activities. This libertarian perspective requires minimal state intervention in citizens' lives and limits equality to formal equality (equality of legal status before the law) rather than substantive equality (equal legal and economic status) (Illuzzi, 2014). Prior *et al.* (1995) use the concepts of *formal citizenship* and *substantive citizenship* to indicate that the former involves the entitlement of rights and duties while the latter goes beyond this, entitling the citizen to receive the required economic support to realise those rights.

Prior *et al.* (1995) point out that both libertarian and social democratic views of citizenship do not enhance political participation beyond a representative system based on regular elections. They instead propose the concept of *associative democracy* as a solution, which consists of devolving power as far as possible to local authorities; these authorities should allow the participation of citizens' associations in their decision-making processes.

Political participation is a widely explored subject in citizenship studies, as the concepts of citizenship and participation are strictly related. As Faulks (2000, p. 4) states, "a key defining characteristic of citizenship, and what differentiates it most from mere subjecthood, is an ethic of participation. Citizenship is an active rather than passive status."

T. H. Marshall, an important contributor to British studies of citizenship, designed an evolutionary pattern of citizenship based on the assertion that it is a "historically variable concept" (cited in Prior *et al.*, 1995, p. 7). He divided citizenship into three categories of rights:

1 "The first category of legal and civil rights enables the individual citizen to participate freely in the life of the community" (Marshall in Prior *et al.*, 1995, p. 7). These rights were established in the seventeenth and eighteenth centuries, at the beginning of the modern state.
2 The second category grants the individual the right of participating in public life through the vote and the democratic representation system. This category was established in Western societies between the nineteenth and twentieth centuries.
3 The third category concerns social and economic rights that developed in the twentieth century, in the post Second World War period, through the establishment of the welfare state. These rights are strictly related to the concept of substantive equality.

Marshall notes that in the early period following the Second World War, Western societies developed the modern welfare state. He argued that a spontaneous evolution of society towards social equality was occurring but, as recent history shows, such evolution has not, in fact, continued, to the extent that scholars now question the compatibility of capitalism with democracy (Merkel, 2014).

Socio-economic equality is considered as a way to reduce social conflict and crime and as a condition to facilitate stronger social cohesion because it can break down the barriers created by social status (Wilkinson and Pickett, 2011). The strongest empirical argument against this kind of equality is the trade-off between efficiency and equality (Okun, 1975). In other words, in a system in which there is a high level of equality, there is usually less motive to be efficient, although this obviously depends on the degree of equality. Justifying some degree of social equality is also a matter of values (see, e.g., DeMarco, 2001) and, in this respect, every attempt to find a widely accepted solution to the debate is likely to be impossible. Instead, research has, in more recent years (Charness and Grosskopf, 2001; O'Connell, 2004; Wilkinson and Pickett, 2011), approached the problem of equality from a different perspective, showing the statistical link between economic equality and concepts such as wellbeing or happiness through empirical research.

Citizenship, the environment and energy

The concept of citizenship has been further elaborated (Smith, 1998; Dobson, 2003) in relation to the environmental problems that society faces.

Dobson (2003) highlights the evolution of citizenship from the liberal and civic republican tradition to the post-cosmopolitan conception that is currently taking shape. Liberalism is recognised as potentially suitable for facilitating the transition towards a sustainable society because of the importance that it attributes to offering individuals and society in general—present and future generations—the ability to choose between several different options.

Furthermore, Dobson (2003) distinguishes between *environmental citizenship* and *ecological citizenship*. Environmental citizenship is grounded in the public sphere and is, in essence, based on the rights and duties attributed to citizens in order to regulate their relationships with the environment in the liberal context of the nation-state. Ecological citizenship is a post-cosmopolitan form of citizenship that is not based on the nation-state: it is non-territorial, yet grounded in both the public and the private spheres and develops from citizenship virtues.

Dobson (2003) regards education as an important means to delivering sustainability with the rationale that a substantial change in society is possible, just by transmitting pro-environmental values to future generations. He contraposes this option with the possibility of using incentives and disincentives, and considers education as best suited to deliver long-lasting change.

Extending citizenship to the field of energy policy, Devine-Wright (2004, 2007) considers that citizenship might deliver the sort of social change that could lead to a low-carbon society. He stresses the importance of citizenship as a status that entails responsibilities which could be enforced but which are subjectively perceived and, in both cases, he holds that they could lead to more sustainable behaviours.

Citizenship and pro-environmental behaviours

The concept of citizenship introduced earlier is an important policy tool for shaping individual agency in order to deliver or protect public goods, especially regarding climate change and the energy transition (Pellegrini-Masini *et al.*, 2019). Public goods, such as the environment or security, are the by-product of social interaction that is governed by formal rules, laws and informal personal and social norms (see, e.g., Putnam, 1993). Therefore, citizenship, both at a formal and at an informal level, can be used to encourage or even impose behaviours that are necessary in order to protect public goods. The specific field of study of environmental citizenship is, in fact, an example of such a use of citizenship in order to address environmental problems (Smith, 1998; Dobson, 2003).

In order to substantially reduce CO_2 emissions, it will, in all likelihood, be necessary to intervene in respect of both the formal rights and duties of citizenship (Pellegrini-Masini *et al.*, 2019) and the practice of citizenship, encouraging its exercise in a pro-environmental way, e.g. participating in renewable energy co-operatives. This kind of participation can be seen as an exercise of active ecological citizenship; it could be encouraged by governments establishing a set of rules and incentives, which would favour this behaviour among other types of citizen participation in renewable energy production.

Theoretical approaches to civic engagement

A number of social science approaches have been used to study different forms of civic engagement and civic participation. What follows is a summary of the relevant research, with a focus on which variables might relate to the acceptability of wind farms and participation in wind farm co-operatives.

Rational choice models

Rational choice models assume that, before acting, citizens evaluate the costs and benefits of the considered action and consequently decide whether or not to act. This model assumes self-interest as the main motive of individual action. While material rewards were originally considered the principal motives, Tyler *et al.* (1986) highlight non-material gains, such as *power* and *prestige*. As Pattie *et al.* (2003) insist, once non-material personal benefits were considered, the literature began to distinguish between *selective* and *collective* benefits. The former category refers to benefits obtained by individuals participating in public life, while the latter refers to benefits available to all as a result of the public activities developed. Three models of participation are presented below, which draw on the rational model but include concepts from social psychology.

General incentives rational action model

In the *general incentives* model outlined by Whiteley and Seyd (1996, cited in Pattie *et al.*, 2003), participation is a function of costs and benefits and *expressive attachments*—to a group or community. These attachments motivate behaviours on behalf of the community. In the general incentives model, a "sense of duty" (Pattie *et al.*, 2003) fosters civic engagement. *Selective benefits* are included and can be divided into three categories: (i) *process benefits*, i.e. benefits deriving from participation in the political process; (ii) *outcome benefits*, privatised outcome benefits; and (iii) *group benefits*, privatised advantages that benefit groups. Additionally, *collective benefits* are those that will be available to the community as a result of collective action and *political efficacy* (the perceived responsiveness of the political system to the expectations of individual citizens), which encourage participation. Finally, this model includes *social norms*, the normative attitudes of family and friends towards participation.

Pattie *et al.* (2003) tested the model against the *social capital* and *civic voluntarism* models (for a description of the models, see pp. 00–00). Their research demonstrated that "low-cost" forms of involvement (e.g. making a donation to an organisation) were more frequent than "high cost" (e.g. participating in a strike) forms of involvement. Their study further showed that *benefits, resources, participation in other organisations* and *mobilisation* were effective in explaining participation.

Participation chain model

In their study of member participation in a co-operative group, Simmons and Birchall (2003) proposed the *participation chain model*, which was based on the *mutual incentives theory* (MIT) developed by Homans (1974, cited in Simmons and Birchall, 2003). MIT joins two theories of motivation, one individualistic and the other collectivistic. The individualistic theory assumes that behaviour is motivated by punishments and rewards. The collectivistic theory developed by Sorokin (1954) considers participation to be motivated by three variables:

1 Shared goals: people express mutual needs that translate into common goals
2 Shared values: people feel a sense of duty to participate as an expression of common values
3 Sense of community: people identify with and care about other people who either live in the same area or are like them in some respect.

    (Sorokin, 1954, cited in Simmons and Birchall, 2003, p. 6).

Individualistic positive incentives include *benefits* and *habits*, while negative incentives include *costs, opportunity costs* and *satiation. Opportunity costs* are determined by the lost opportunities that arise from declining to participate in a particular activity. Benefits are subdivided into *external* and *internal* benefits; external benefits are material/tangible advantages, while internal benefits are subjectively perceived advantages.

However, Simmons and Birchall (2003) regarded MIT as not sufficiently inclusive of all the variables involved in influencing participation. Therefore, the authors included the categorisation suggested by Whiteley and Seyd (1996, cited in Pattie *et al.*, 2003), a scheme that included incentive-based explanations as *demand-side* models of participation as well as *supply-side* models. These models led to the inclusion of the variables *personal resources* and *mobilisation factors.* Subsequently, Simmons and Birchall (2003) expanded their MIT to include these factors as a stage of a multilevel *participation chain:* the first level consists of *resources,* including time, money, skills and confidence; the second level involves the *mobilisation* of resources. Mobilisation is comprised of three elements: *issues, opportunities* and *recruitment efforts.* Finally, the last stage is *motivations* (as originally outlined in the MIT). The term *issues* concerns the importance that the object of participation assumes for participants, a variable whose importance has been remarked by others (Lowndes *et al.,* 2001). *Opportunities* are the presence of good quality opportunities to participate. *Recruitment efforts* are passive or active recruitment activities. By passive, the authors mean that the opportunity is made public without directly asking potential participants; active recruitment will involve this direct asking.

The authors tested the model in a study regarding participation in the governance of a UK co-operative group (Birchall and Simmons, 2004). Members of the co-operative group were sensitive to the benefits that the experience of participating provided to them, although internal benefits (e.g. valuable learning experience) were mostly significant for respondents rather than external benefits (e.g. financial reward). Interestingly, collectivistic incentives were the most important individual benefits: this result is probably related to the co-operative specificity of an organisation highly motivated by a certain set of values (social justice, egalitarianism). Finally, mobilisation was significant in eliciting participation, particularly skills derived from previous experience and face-to-face recruitment efforts.

A rational choice model for environmental collective activism

Lubell (2002) presented a rational model of collective action for environmental activism, in which he adapted the collective interest model, a model of collective action derived from the work of Olson (1965). In his model, Lubell summarised the variables influencing environmental activism in the following equation:

$$EV = [(Pg + Pi) * V] - C + B$$

In this equation, EV is environmental activism, the expected value of participation; Pg is group efficacy; Pi is personal efficacy; V is the value of the collective good; C is the selective cost of participation; and B is the selective benefit of participation (Lubell, 2002, p. 433). Although this model includes some psychological variables (for example, personal efficacy), it lacks a comprehensive view of the variables involved in the personal choice to engage in *environmentally responsible behaviours* (ERBs), such as attitudinal variables or social pressure.

Justice-based models

Tyler *et al.* (1986) outlined a *fairness-based psychological model* or *justice-based model.* The authors distinguish between a procedural justice-based model and a distributive justice-based model, although these models are viewed as operating simultaneously. Their basic assumption is that citizens judge the fairness of a certain process produced by policies as well as the fairness of its outcomes. In the case of distributive justice, the focus of the judgement is on the outcome of the process. In the case of procedural justice, the evaluation is on the process. Tyler *et al.* suggest that governments could retain the support of citizens in the presence of a process perceived as fair, even if the outcome would penalise citizens: "Similarly a justice based perspective suggests that government might be able to restrict citizens and limit the benefits they receive in a future era of scarcity without losing support from the public if they do so fairly" (Tyler *et al.*, 1986, p. 976). Tyler *et al.* (1986) considered data collected in 14 different studies; in these studies, judgements about gain and loss (*rational choice*) were compared with justice-based evaluations in their influence on political evaluations, voting behaviours and non-voting behaviours (e.g. writing to a member of Congress). When the two different categories of judgements were compared, the importance of justice-based judgements in influencing the dependent variables became clear.

Socio-psychological models

Some theoretical frameworks of citizen participation stress the importance of socio-psychological variables. The discussion below describes them and then highlights commonalities which characterise them.

The dual pathway model of socio-political participation

Stürmer and Kampmeier (2003, p. 108) conceived a framework that considers both incentives and costs and collective identification processes: "In line with traditional social psychological approaches, one pathway is based on the calculation of costs and benefits; the other is based on collective identification processes." The incentives are at the individual level, while *community identification* is regarded separately as an antecedent of participation. In Stürmer and Kampmeier's empirical studies, identification with the local community was predictive of participation in community volunteerism and in local political and social initiatives.

The social-cognitive approach of Albert Bandura

Although Bandura (2000, 2001, 2002) does not offer an explicit model of participation, he uses concepts from *social cognitive theory* to point to a few motives that lead people to participate. Social cognitive theory distinguishes between three different forms of agency: (i) *personal*, (ii) *proxy* and (iii) *collective*. Personal agency is at the individual level and is the most widely studied by social cognitive theory, while proxy agency is used by individuals when they lack the skills or time to develop the required skills or the time to act on their own behalf and therefore act through a third person. In the case of collective agency, people feel the need to act together to achieve common goals that require interdependent efforts (Bandura, 2000). At an individual level, perceived self-efficacy is the perception that an individual holds of his or her suitability to engage successfully in action, while at a collective level, "[p]eople's shared beliefs in their collective power to produce desired results are a key ingredient" (Bandura, 2000, p. 75). Hence, *perceived collective efficacy* is a group-level property that is not the sum of the group members' individual self-efficacy (Bandura, 2000).

Thus, individual self-efficacy can have a role in a person's deciding to join a group, and the perceived collective efficacy of the group can make the group seen more attractive to non-members so that they consider joining it. As Bandura (2000) points out, collective efficacy influences group members' efforts and their resilience when faced with a lack of results. It is worth noting that *confidence* plays an important role in the earlier introduced participation chain of Simmons and Birchall (2003). Confidence appears to be similar to perceived individual self-efficacy. It is included (Simmons and Birchall, 2003) within the set of personal resources along with time, money and skills, although, as Bandura points out, resources can also influence self-efficacy or confidence: "Economic conditions, socioeconomic status, and family structure affect behaviour through their impact on people's sense of efficacy, aspirations, and affective self-regulatory variables rather than directly" (Bandura, 2000, p. 76).

The value-belief-norm theory (VBN) of support for social movements

Stern *et al*. (1999) outline a theory of support for environmental social movements that is an attempt to explain four behaviours: demonstrating, low committed citizenship behaviours, policy support and changes in personal- and private-sphere behaviours. The theory presents a casual chain linking *values* to the *new ecological paradigm theory* (NEP), *awareness of consequences* (AC), *ascription of responsibilities* (AR) and *normative beliefs*; these norms should imply the adoption of environmentally responsible behaviours (ERBs). Each of these variables influences the subsequent ones down the chain. Personal norms are ultimately influenced by the awareness of adverse consequences (AC) on objects that people value, and by the belief that personal behaviour can reduce this threat (AR).

Stern (2000) pointed out that personal capabilities, in addition to attitudes, play a fundamental role in shaping pro-environmental behaviours. This was also suggested by Stern *et al*. (1999): e.g. *demonstrating* was not explained in VBN theory, but it was found to be positively related to age and income. VBN theory, therefore, provides a useful insight in accounting for cognitive processes that motivate participatory behaviours, although it does not seem sufficient in explaining these behaviours completely.

Social capital

Putnam (1993) was moved to investigate the positive relation between social capital and democracy, economic development and societal wellbeing. *Social capital theory* provides some useful insights into the relation between social structure and civic engagement. Putnam considers modern societies and their social culture using a historical perspective; he compares societies ruled by vertical organisations (scarcely democratic), in which social trust and social networks are weak, with societies in which horizontal democratic organisations— such as mass political parties, co-operatives and associations—are well developed. Putnam (1993, p. 88) regards self-interest and altruism as only apparently opposed; he thinks that "enlightened" self-interest motivates people to pursue collective benefits.

It seems that this convergence of individual and collective interests is similar to Bandura's (2000) description of collective agency, and it appears to be accounted for also in the participation chain of Simmons and Birchall (2003) where, between collective incentives, we find *shared goals*. Putnam (1993) appears to affirm a mutual reinforcement between *interaction, trust* and *reciprocity*, but does not provide a specific theory of civic engagement that could explain further the presence or the lack of such social features. Finally, his historical perspective leads Putnam (1993, p. 183) to affirm that civic community "has deep historical roots", and to conclude that change is a very slow process that does not occur merely from a willingness to reform.

Civic voluntarism: a socio-economic approach

Civic voluntarism is a socio-economic model of participation which holds that "the better educated, more affluent and more middle class people are, the more likely they are to participate" (Pattie *et al.*, 2003, p. 445).

Socio-economic status (SES: education, income and occupation) is regarded as a well-demonstrated variable influencing participation (Verba *et al.*, 1995). Building on this evidence, Brady *et al.* (1995) propose to substitute SES with a resource model that includes *time*, (availability of free time), *civic skills* (communication and organisational capacities), *money* (household income) and *psychological engagement* in politics, an attitudinal measure. In their opinion, these resources could refine the predictive efficacy of SES, while maintaining the same rationale.

Finally, in this framework, mobilisation is an important variable for determining participation: in particular, asking people to participate and stressing the importance of participation are considered the main mobilisation actions that foster participation. Of these variables, Brady *et al.* (1995) stress the importance of resources, which, when compared with psychological engagement, are more easily measurable and therefore less disputable in academic debates.

Discursive psychology approach: citizenship in practice

Barnes *et al.* (2004) use discursive psychology to study citizens' opinions on the specific matter of New Age travellers. Despite the specificity of the case under consideration, this approach uses the tools of discursive psychology and content analysis to describe how citizens build their identities in relation to their communities and specific areas of residence. Citizens represent to themselves and to others their identities, and the identities of others, in order to claim rights and to achieve goals. The apparent objective of citizens' action is the pursuit of personal or group interests, but the discursive psychology approach does not investigate how cognitive processes shape interests and consequent actions; in fact, this approach rejects the assumption that discourse is a mirror of cognitive processes (Potter, 1996).

Final remarks about the theories of civic engagement considered

A few generalizations that can be drawn, based on the discussion above.

1 Most authors (Putnam, 1993; Bandura, 2000; Lubell, 2002; Pattie *et al.*, 2003; Simmons and Birchall, 2003; Stürmer and Kampmeier, 2003) consider incentives or interests as motivating participation, and some (Lubell, 2002; Pattie *et al.*, 2003; Simmons and Birchall, 2003) highlight the confluence of individual (or selective) incentives with collectivistic incentives in determining participation.
2 Processes of identification or attachment towards a group or community that could benefit from civic engagement were considered by several authors to encourage participation (Pattie *et al.*, 2003; Simmons and Birchall, 2003; Stürmer and Kampmeier, 2003).
3 Perceived efficacy is identified as fostering participation (Bandura, 2000; Lubell, 2002; Simmons and Birchall, 2003).
4 Several authors (Verba *et al.*, 1995; Stern, 2000; Simmons and Birchall, 2003) consider personal resources to positively influence participation either directly or indirectly or through perceived efficacy (Bandura, 2000).
5 *Responsibility* is included in two of the models: in the *value-belief-norm theory* (Stern *et al.*, 1999) and in the *general incentives rational action model* (Pattie *et al.*, 2003).

Citizen participation in the context of sustainable energy developments

When considering participation in the context of renewable energy policy and the social acceptability of renewable energy developments, it is necessary to draw a line between what could be termed *passive* and *active* involvement.

Passive involvement would be limited to simple acceptance of an energy policy or of new renewable developments across the country, including acceptance of local installations of renewable generators. Such passive involvement is obviously very different from an active engagement in promoting renewables. This distinction appears in the work of Batel *et al.* (2013) who use the same line of reasoning to distinguish between *acceptance* and *support*. Therefore, it is likely that active and passive involvement (or support and acceptance) would be produced by different determinants. In the case of acceptance (or acceptability)—in other words, passive involvement—there is a sense of responsibility towards local, national and global communities and/or future generations; a perception of the procedural justice of a government's policies; and a perception of distributive justice in terms of outcomes that could play a role in determining consensus towards an energy policy promoting change. A sense of responsibility could be fostered by a sense of identification or attachment towards the community at the local, national and global levels. As Stürmer and Kampmeier (2003, p. 107) point out in their discussion of *collective decisions*, "[g]roup members perceive these decisions and norms as socially valid and meaningful. This, in turn, fosters an inner commitment to act accordingly and behave as a 'good' (i.e. responsible) group member."

Personal resources (such as time, income, skills, education) could instead play a different role in regard to passive acceptability rather than in active engagement. In fact, passive acceptance of policies could be related to low income and lack of education: people with little education or fewer skills may refrain from elaborating a personal opinion and passively accept their government's action out of a sense of responsibility towards the nation or, more simply, due to a lack of confidence.

Therefore, active engagement in organisations, associations and co-operatives or actively supporting renewable energy developments may, in part, have a different set of determinants. If identification processes are important, other determinants such as individual incentives, self-efficacy and personal resources (income, skills, education and time) come into play. At the same time, mobilisation stimulated by local or national authorities or groups interested in expanding their structures could also play a role in fostering active participation.

Environmentally responsible behaviour and its determinants

Environmentally responsible behaviour (ERB) or *pro-environmental behaviour* is a behaviour that consciously limits negative consequences on the environment (Kollmuss and Agyeman, 2002). In contrast, the label "environmental significant behaviour" does not have a positive or negative connotation, it simply refers to a behaviour that has an impact on the environment (Stern, 2000).

While the literature on ERB is vast, there is a lack of consensus regarding its determinants. As Barr (2003) points out, different variables influence ERBs, depending on the behaviour itself, an assertion confirmed by a relatively recent meta-analysis (Osbaldiston and Schott, 2011). De Young (2000, p. 510) makes the same claim, stating that, "empirical evidence has emerged supporting the idea that ERB has multiple antecedents", and that different behaviours may have different determinants. However, these assumptions do not mean that it is impossible to consider a single behaviour or a set of similar behaviours as shaped by a group of variables that influence them. Research on ERBs makes it clear that some early assumptions were rejected by the empirical tests conducted, particularly the assumption that widespread information on environmental matters would simply, as a matter of course, induce people to adopt ERBs (Kollmuss and Agyeman, 2002; Barr, 2003) or that altruism-based models (Stern *et al.*, 1993) could explain why ERBs take place instead of behaviours motivated by self-interest.

It has been suggested (Stern, 2000; Barr, 2003) that situational barriers play an important role in preventing ERBs. More specifically, Barr (2003) shows how in relation to a behaviour like recycling, variables such as available facilities and collection schemes play an important role in determining ERBs. However, environmental values, a sense of responsibility towards the environment—what Stern (2000) and Barr (2003) term "*environmental citizenship*"—as well as previous personal experiences and demographic variables, had an important role in fostering waste minimisation, an ERB that does not require specific facilities.

If barriers play a role in environmental behaviours, this could mean that citizens will avoid behaviours that require a great deal of personal sacrifice, although some highly-committed environmentalists might carry them out anyway. It might be that for many individuals, the decision to engage in ERBs involves a personal assessment of the costs and benefits of every single action before engaging in it (Diekmann and Preisendörfer, 2003). In this regard, Diekmann and Preisendörfer integrate a rational choice model with *environmental attitude* research. The authors examined the influence of attitudes in high-cost versus low-cost situations and proposed the "low-cost hypothesis":

Environmental concern influences ecological behaviour primarily in situations and under conditions connected with low-costs and little inconvenience for individual actors. The lower the pressure of costs in a situation, the easier it is for actors to transform their attitudes into corresponding behavior. If costs are high, environmental concern does not help overcome one's reservations, and there will be few or no effects of environmental attitudes.

(Diekmann and Preisendörfer, 2003, p. 443)

The low-cost hypothesis does not exclusively consider economic costs. In considering rational choice models, Diekmann and Preisendörfer (2003) criticise their inability to explain individual choice in low-cost situations; they maintain that these models work to explain individual choice in high-cost situations only, and that below a certain cost threshold, attitudes have a major influence on behaviour. As noted by other authors (Stern, 2000; Kollmuss and Agyeman, 2002) individuals are more likely to participate in ERBs when they do not require high personal costs (money and/or time). This view is therefore consistent with the approach of Kaplan (2000) and De Young (2000) who hold the opinion that altruism alone cannot explain ERBs, especially in the long run. These authors share the perspective that self-interest is a major driver of human behaviour and that, despite the fact that altruism could offer a starting motivation for engaging in an ERB, in the long term, self-interest—which De Young (2000, p. 515) identifies as a form of "intrinsic satisfaction"—could keep a subject actively committed to a certain behaviour. De Young (2000, p. 521) holds the opinion that competence is a key aspect of intrinsic satisfaction: "People find unpleasant and thus avoid situations in which they cannot advance or utilize their competence."

Final remarks about the literature on environmentally responsible behaviour

The following conclusions can be drawn from the literature that has been reviewed on ERBs:

1 ERBs have different determinants depending on the kind of behaviour that we consider.
2 Different behaviours could share the same determinants, but the relative effect on influencing behaviours would vary from one kind of behaviour to another.
3 Information and environmental awareness alone are not able to predict ERBs.
4 Attitudes alone are insufficient to predict ERBs, but they have more predictive power when the considered behaviours do not incur costs from situational barriers and strong disadvantages (for example, money and time costs).

References

Bandura, A. 2000. Exercise of human agency through collective efficacy. *Current Directions in Psychological Science*, 9, 75–78.

Bandura, A. 2001. Social cognitive theory: An agentic perspective. *Annual Review of Psychology*, 52, 1–26.

Bandura, A. 2002. Selective moral disengagement in the exercise of moral agency. *Journal of Moral Education*, 31, 101–119.

Barnes, R., Auburn, T. and Lea, S. 2004. Citizenship in practice. *British Journal of Social Psychology*, 43, 187–206.

Barr, S. 2003. Strategies for sustainability: Citizens and responsible environmental behaviour. *Area*, 35, 227–240.

Batel, S., Devine-Wright, P. and Tangeland, T. 2013. Social acceptance of low carbon energy and associated infrastructures: A critical discussion. *Energy Policy*, 58, 1–5.

Batel, S., Devine-Wright, P., Wold, L., Egeland, H., Jacobsen, G. and Aas, O. 2015. The role of (de-)essentialisation within siting conflicts: An interdisciplinary approach. *Journal of Environmental Psychology*, 44, 149–159.

Birchall, J. and Simmons, R. 2004. What motivates members to participate in co-operative and mutual businesses? *Annals of Public and Cooperative Economics*, 75, 465–495.

Blunkett, D. 2003. *Civil renewal: A new agenda*. Home Office, London. Available at: www.homeoffice.gov.uk/docs2/civilrennewagenda.pdf.

Brady, H. E., Verba, S. and Schlozman, K. L. 1995. Beyond SES: A resource model of political participation. *American Political Science Review*, 89, 271–294.

Cambridge Dictionary. n.d. *Acceptance* [Online]. Cambridge Advanced Learner's Dictionary & Thesaurus. Available at: https://dictionary.cambridge.org/dictionary/english/acceptance [Accessed 5 April 2018].

Charness, G. and Grosskopf, B. 2001. Relative payoffs and happiness: An experimental study. *Journal of Economic Behavior & Organization*, 45, 301–328.

De Young, R. 2000. New ways to promote proenvironmental behavior: Expanding and evaluating motives for environmentally responsible behavior. *Journal of Social Issues*, 56, 509–526.

DECC. 2011. *UK renewable energy roadmap*. Department of Energy & Climate Change, London Available at: www.gov.uk/.../2167-uk-renewable-energy-roadmap.pdf.

DECC. 2014. *Community energy strategy*. Department of Energy & Climate Change, London. Available at: www.gov.uk/government/uploads/system/uploads/attachment_data/file/275163/20140126Community_Energy_Strategy.pdf.

DECC. 2015. *Community energy strategy update*. Department of Energy & Climate Change, London. Available at: www.gov.uk/government/uploads/system/uploads/attachment_data/file/414446/CESU_FINAL.pdf.

DeMarco, J. P. 2001. Substantive equality: A basic value. *Journal of Social Philosophy*, 32, 197–206.

Department for Business, Energy & Industrial Strategy. 2019. *The UK's draft integrated National Energy and Climate Plan (NECP)*. Department for Business, Energy & Industrial Strategy, London. Available at: https://ec.europa.eu/energy/sites/ener/files/documents/unitedkingdom_draftnecp.pdf.

Devine-Wright, P. 2004. Towards zero-carbon: Citizenship, responsibility and the public acceptability of sustainable energy technologies. In: Buckle, C. (ed.) *Proceedings of Conference C81 of the Solar Energy Society*, UK Section of the International Solar Energy Society, 21 September 2004, London.

Devine-Wright, P. 2007. Energy citizenship: Psychological aspects of evolution in sustainable energy technologies. In: Murphy, J. (ed.) *Governing technology for sustainability*. London: Earthscan.

Diekmann, A. and Preisendörfer, P. 2003. Green and greenback: The behavioral effects of environmental attitudes in low-cost and high-cost situations. *Rationality and Society*, 15, 441–472.

Dinham, A. 2010. *Active citizenship and the big society*. Faiths & Civil Society Unit, Goldsmiths, University of London. Available at: www.gold.ac.uk/media/documents-by-section/departments/research-centres-and-units/research-centres/centre-for-lifelong-learning/Active-Citizenship-and-the-Big-Society.doc.

Dobson, A. 2003. *Citizenship and the environment*. Oxford: Oxford University Press.

DTI and DEFRA. 2003. *Our energy future: Creating a low carbon economy*. Department of Trade and Industry, London. Available at: www.dti.gov.uk/energy/whitepaper/ourenergyfuture.pdf.

Faulks, K. 2000. *Citizenship*. New York: Routledge.

HM Government. 2009. *The UK renewable energy strategy*. The Stationery Office, London. Available at: www.official-documents.gov.uk/document/cm76/7686/7686.pdf.

Homans, G. 1974. *Social behaviour: Its elementary forms* (2nd edn.). New York: Harcourt Brace Jovanovich.

Illuzzi, M. J. 2014. Equality. In: Gibbons, M. T., Coole, D., Ellis, E. and Ferguson, K. (eds.) *The encyclopedia of political thought*. Chichester, UK: Wiley-Blackwell.

Kaplan, S. 2000. New ways to promote proenvironmental behavior: Human nature and environmentally responsible behavior. *Journal of Social Issues*, 56, 491–508.

Kollmuss, A. and Agyeman, J. 2002. Mind the gap: Why do people act environmentally and what are the barriers to pro-environmental behavior? *Environmental Education Research*, 8, 239–260.

Lipp, J. and McMurtry, J. 2015. *Benefits of renewable energy co-operatives: Summary of literature review from the Measuring the Co-operative Difference Research Network.* Measuring the Co-operative Difference Research Network, Canada. Available at: http://hdl.handle.net/10587/1608.

Local Energy Scotland. 2015. *Good practice principles for shared ownership of onshore renewable energy developments.* Energy Saving Trust, Edinburgh. Available at: www.localenergyscotland.org/media/79714/Shared-Ownership-Good-Practice-Principles.pdf.

Lovins, A. B. 1977. *Soft energy paths: Toward a durable peace.* Cambridge, MA: Ballinger Publishing and Friends of the Earth International.

Lowndes, V., Pratchett, L. and Stoker, G. 2001. Trends in public participation: Part 2— citizens' perspectives. *Public Administration,* 79, 445–455.

Lubell, M. 2002. Environmental activism as collective action. *Environment and Behavior,* 34, 431–454.

McLean, I. and McMillan, A. 2009. *The concise Oxford dictionary of politics.* Oxford: Oxford University Press.

Merkel, W. 2014. Is capitalism compatible with democracy? *Zeitschrift für Vergleichende Politikwissenschaft,* 8, 109–128.

O'Connell, M. 2004. Fairly satisfied: Economic equality, wealth and satisfaction. *Journal of Economic Psychology,* 25, 297–305.

Okun, A. M. 1975. *Equality and efficiency: The big tradeoff.* Washington, DC: The Brookings Institution.

Olson, M. 1965. *The Logic of collective action: Public goods and the theory of groups* (revised edn.). London: Harvard University Press.

Osbaldiston, R. and Schott, J. P. 2011. Environmental sustainability and behavioral science: Meta-analysis of proenvironmental behavior experiments. *Environment and Behavior,* 44, 257–299.

Patterson, W. 1999. *Transforming electricity: The coming generation of change.* Abingdon, UK: Earthscan.

Pattie, C., Seyd, P. and Whiteley, P. 2003. Citizenship and civic engagement: Attitudes and behaviour in Britain. *Political Studies,* 51, 443–468.

Pellegrini-Masini, G., Corvino, F. and Pirni, A. 2019. Climate justice in practice: Adapting democratic institutions for environmental citizenship. In: Harris, P. G. (ed.) *A research agenda for climate justice.* London: Edward Elgar Publishing.

Potter, J. 1996. *Representing reality: Discourse, rhetoric and social construction.* London: Sage.

Prior, D., Stewart, J. and Walsh, K. 1995. *Citizenship: Rights, community and participation.* London: Pitman.

Putnam, R. 1993. *Making democracy work: Civic traditions in modern Italy.* Princeton, NJ: Princeton University Press.

Reeve, A. 2009. Citizenship. In: McLean, I. and McMillan, A. (eds.) *The concise Oxford dictionary of politics* (3rd edn.). Oxford: Oxford University Press.

Rifkin, J. 2002. *The hydrogen economy: The creation of the worldwide energy web and the redistribution of power on Earth.* Cambridge: Polity Press.

Scottish Government. 2011. *2020 routemap for renewable energy in Scotland.* APS Group Scotland, Edinburgh. Available at: www.gov.scot/Resource/Doc/917/0118802.pdf.

Simmons, R. and Birchall, J. 2003. Bringing citizens back into public services: Strengthening the "Participation Chain". ECPR Joint Sessions, Edinburgh, 28 March–2 April 2003.

Smith, M. J. 1998. *Ecologism: Towards ecological citizenship*. Buckingham, UK: Open University Press.

Sorokin, P. 1954. *The ways and power of love*. Boston, MA: Beacon Press.

Stern, P. C. 2000. New environmental theories: Toward a coherent theory of environmentally significant behavior. *Journal of Social Issues*, 56, 407–424.

Stern, P. C., Dietz, T. and Kalof, L. 1993. Value orientations, gender, and environmental concern. *Environment and Behavior*, 25, 322–348.

Stern, P. C., Dietz, T., Abel, T. D., Guagnano, G. A. and Kalof, L. 1999. A value-belief-norm theory of support for social movements: The case of environmentalism. *Human Ecology Review*, 6, 81–97.

Stürmer, S. and Kampmeier, C. 2003. Active citizenship: The role of community identification in community volunteerism and local participation. *Psychologica Belgica*, 43, 103–122.

Szarka, J. 2007. *Wind power in Europe: Politics, business and society*, Basingstoke, UK: Palgrave Macmillan.

The CIC Project Team. 2004. *Community interest companies: An introduction to community interest companies*. Department of Trade and Industry, London. Available at: www.dti.gov.uk/cics/pdfs/cicfactsheet1.pdf.

Tyler, T. R., Rasinski, K. A. and Griffin, E. 1986. Alternative images of the citizen: Implications for public policy. *American Psychologist*, 41, 970–978.

Verba, S., Schlozman, K. L. and Brady, H. E. 1995. *Voice and equality: Civic voluntarism in American politics*. Cambridge, MA: Harvard University Press.

Whiteley, P. and Seyd, P. 1996. Rationality and party activism: Encompassing tests of alternative models of political participation. *European Journal of Political Research*, 29, 215–234.

Wilkinson, R. and Pickett, K. 2011. *The spirit level: Why greater equality makes societies stronger*. New York: Bloomsbury Press.

Wüstenhagen, R., Wolsink, M. And Bürer, M. J. 2007. Social acceptance of renewable energy innovation: An introduction to the concept. *Energy Policy*, 35, 2683–2691.

2 Wind energy acceptability—what, how and when

All the variables at stake

The following chapter surveys a considerable amount of research that has been published on the social acceptability of wind farms, their perceived benefits to the communities near where they are located and wind farm co-operatives. In particular, it focuses on the factors that the literature has found to influence opposition to or support for wind farms, and concludes with a discussion of how this study builds upon previous work.

Opposition to wind energy

While a national 2010 survey revealed that popular support for wind energy had reached 82 per cent in the UK (Spence *et al.*, 2010, p. 3), support may have been weaker in the specific areas where developments were proposed (Devine-Wright, 2005a).

Local opposition to wind energy has been explained in terms of three categories of factors, which echo the classification of factors that influence environmentally significant behaviours proposed by Stern (2000) and adapted by Devine-Wright (2008). They are: (i) *attitudinal factors*, comprised of perceived local costs and benefits of local relevance, perceived non-local costs and benefits, place attachment and environmental attitudes; (ii) *personal factors*, comprised of knowledge about wind energy, affluence/deprivation and proximity of residence to a wind farm; and (iii) *contextual factors*, comprised of procedural fairness, trust towards the proponents and participation.

Choosing which factors belong in each of the categories is a matter of some contention. Ultimately, most of the factors can be considered attitudinal, in the sense that their perception is generally subjective. For example, different people might have a range of opinions about issues such as the information that a developer distributed to local residents. Moreover, a single factor, such as "proximity of residence to wind farm" could be considered a "contextual factor" rather than a "personal factor". Therefore, the classifications utilised here differ in some ways from those proposed by Stern (2000) and Devine-Wright (2008), not only in terms of the choice of the single factors but also in the placement of a factor in one category rather than another.

Perceived local impacts

Factors of local importance have traditionally explained much of the opposition to wind farms. As Agterbosch *et al.* (2009, p. 400) state: "The mismatch between the local common interest and the external private or global environmental interest contributes to the risk of local social resistance." The environmental benefits of wind farms accrue globally (i.e. diminished carbon emissions), while their economic advantages are entirely or mainly delivered outside of the local community (to developers). Therefore, the local community that surrounds the site bears the impact of the development on its local environment and does not receive substantial benefits.

Cass *et al.*'s research (2010) confirms this. They sent a questionnaire to local residents living close to ten renewable energy projects in the UK. The data from their survey, drawn from 2911 respondents, demonstrated a strong positive correlation between perceived personal benefits and support for the project. Therefore, the more respondents perceived that the energy project would benefit them or their local area, the more likely they were to offer their support. Perceived local and personal benefits and impacts were also highly correlated (+0.85). Cass *et al.* (2010) also carried out a linear multiple regression analysis to compare the effects of benefits/impacts with other variables that might influence project support, such as attitudes to the technology and perceptions of trust. They found that the perception of benefits was the single most important determinant of project support, a factor even more important than "general beliefs about the technology sector", "beliefs about the developer's engagement practices", "trust in the developer" and "perceived fairness of the planning procedures".

Attitudinal factors

Attitudinal factors included the perceived costs and benefits (local and non-local) of a wind farm, the perceived trustworthiness of wind farm proponents, place attachment and environmental attitudes. All of these factors are included in the "attitudinal factors" category under the classification proposed by Devine-Wright (2008). This research attempted a detailed sub-classification of the perceived local and non-local impacts, something that, in the meantime, other authors have proposed (Perlaviciute and Steg, 2014).

Perceived local costs and benefits

Perceived local costs and benefits include, but are not limited to, the following factors: local economic impact of the wind farm; the visual impact, the auditory impact and related noise pollution; and the perceived impact on the health of the local population.

Local economic factors and community benefits

Economic interests are the main driver for wind energy developers, but they also influence the local community's opinions of the development.

A range of local economic issues can affect residents' opinions. On the one hand, they may be wary of the development's potential impact on local property values or its effects on the local tourism industry, if there is one. On the other hand, they may believe that a local wind farm could bring advantages, such as local jobs or community benefits (donations or revenue to the community provided by the developer). Finally, local ownership schemes can influence community attitudes because they allow local residents to profit from the revenue generated by the wind farm.

In this regard, Agterbosch *et al.* (2009) conducted a multiple case study survey in the Netherlands, conducting semi-structured interviews with wind farm stakeholders and content analysis of written material. Looking at the municipality of Zeewolde, which alone accounts for 40 per cent of Dutch wind energy, they concluded that shared economic interest between the developer and the local community, along with procedural fairness, can substantially diminish local opposition.

Jones and Eiser (2009) also looked into the social acceptability of wind farms. They examined four proposed wind farms in suburban areas of Sheffield (England) adjacent to the chosen sites (target areas), and four comparison areas in similar suburban areas of Sheffield at a greater distance from the chosen wind farm sites; 1200 questionnaires were distributed on a door-to-door basis, of which 843 were returned. Several hierarchical regression analyses were carried out that analysed the influence of groups of variables on support/opposition, while controlling for "general attitude to wind energy". One of these groups was economic variables, and four were significantly related to support in the target areas: "general economic benefit" (0.30), "community trust fund" (0.10), "opportunity to invest" (0.13) and "cheaper electricity" (0.13). "General attitude to wind" was the most influential (0.32). In the comparison areas, the only two significant variables in addition to "general attitude to wind" (0.50) were "general economic benefit" (0.15) and "employment opportunity" (0.09). Therefore, the regression analysis revealed that "general attitude to wind" was the most influential variable in determining support/opposition but, in the case of the target areas, it was very closely followed by "general economic benefit". This demonstrates that respondents in the target areas were sensitive to the potential economic advantages deriving from the proposed wind farms. In the comparison areas, the difference in the standardised beta coefficients between "general attitude to wind" and economic variables demonstrates that respondents were less influenced by these variables in their support for the proposed wind farms. This was possibly because the comparison areas' distance from the wind farm sites led respondents to believe that they would be less influenced by possible economic impacts.

Furthermore, in a survey of local acceptance of a proposed off-shore wind farm on Tunes Plateau (off the coast of Northern Ireland), Ellis *et al.* (2007) found that anti-wind rhetoric was tied to respondents' economic motives. The authors used q-methodology, which employs statistical analysis to analyse qualitative data and establish patterns between different discourses (Barry and

Proops, 1999). They determined four factors that jointly explained 62 per cent of the total variance in the data generated by interviews with self-declared opponents. Of these four factors, "anti-wind power-local resister", a cluster of opinions dominated by a distrust of wind energy as a viable and effective means of electricity production and the perception that it will have a negative economic impact on the local community, accounted for 17 per cent of the total variance. Two other factors, the "anti-developer—pragmatic localist" and the "economic sceptic-siting compromiser", accounted for 14 per cent and 10 per cent, respectively, of the total variance. Respondents pertaining to these categories shared one core belief: that "the scheme will result in a negative local economic impact" (Ellis *et al.*, 2007, p. 527).

Toke (2005b) also attributes great importance to the economic impacts of a wind farm. He carried out a regression analysis of the planning data regarding 51 wind farm cases in England and Wales. He found that there was a statistically significant relationship between the independent variables "opinion of the parish council", "planning officer's recommendation" and "opinion of the countryside protection group" and the dependent variable of "decision of the local planning authority". Toke discusses his results in the context of further research data gathered through interviews and written material obtained from local authorities, the Department of Trade and Industry (DTI), the Office of the Deputy Prime Minister (ODPM) and the British Wind Energy Association (BWEA). He concludes that, "[t]he dominant local perception of the economic impact of the scheme on the locality is also an especially crucial variable" (Toke, 2005b, p. 1539). In discussing the specific perceived economic impacts, Toke mentions the impact on the tourism industry: whether negative or positive, the perception of the wind farm as an element capable of restoring a positive economic image of an otherwise failing local economy, and the presumed capacity of wind energy to create local jobs. He suggests (Toke, 2005b), moreover, that subjects may perceive the economic potential of wind farms differently depending on their social class. For example, while farmers might view wind farms as an opportunity, affluent middle-class residents may favour the preservation of landscape views and, ultimately, their property values.

The next section details the specific economic factors that influence local residents' opinions of wind farm developments.

Local property values

A number of studies have assessed the effect of wind farms on property values (Sterzinger *et al.*, 2003; Sims and Dent, 2007; Sims et al., 2008). They have found either no evidence of decreased property values (in the case of the American study by Sterzinger *et al.*) or tenuous, non-conclusive support for decreased property values (in the case of the UK studies conducted by Sims *et al.*). (The details of these studies on the real effect of wind farms on local property prices will not be discussed here because the main interest is in the locally-perceived risk of devaluation of properties, which is a patently different, though related, phenomena from the actual depreciation of properties once the wind farm has been built.)

Jones and Eiser's (2009) previously introduced study dealt with the topic of perceived risk of property devaluation. In their study, the authors carried out a hierarchical regression analysis of the variables that are perceived as local disadvantages of wind farms, while controlling for a variable called "general attitude to wind". Their results showed that for residents in the target areas (individuals living adjacent to proposed wind farm sites), "general attitude to wind" (0.28), "spoil the landscape" (–0.23), "lower house prices" (–0.14) and "general unwanted change" (–0.33) were statistically significant factors influencing support or opposition to the development. Other variables, including "noise" and "hazardous to health", were not found to be statistically significant. While it is not fully explained what the "general unwanted change" variable exactly represents,[1] Jones and Eiser's study (2009, p. 4611) nevertheless indicated that residents were most wary of the proposed development's impact on the landscape, followed immediately by its potential to lower house prices.

Firestone and Kempton's (2007) survey of 504 residents near a proposed wind farm off Cape Cod (Massachusetts, USA) indicated that the odds of supporting a proposed wind farm decreased by 88 per cent if respondents identified "property values" as one of the three main issues related to the proposal. In their study, only 14 per cent of opponents identified property values as one of the three most important issues affecting their opposition. Nevertheless, 48 per cent of all respondents believed that the development would have had a negative impact on "property values".

Taken together, the two aforementioned studies suggest that a perceived negative impact on property prices negatively influences a proposed wind farm's acceptability. However, these concerns do not appear to be widespread.

Effect on tourism

Another feared negative economic effect of the presence of wind farms is that on local tourism. Toke (2005b, p. 1538) found that in a quarter of the 51 wind farm planning cases he studied, the fears of negative repercussions on local tourism were a "significant issue", and all of these cases included areas with a significant number of tourism-related economic activities.

On the other hand, in a comparative study of two touristic areas in the Czech Republic, one with a proposed wind farm and another with a built wind farm, Frantál and Kunc (2011), found that local residents employed in the local tourism industry did not have strong negative views of the wind farm's influence on tourism in either of the areas surveyed. Twelve per cent of respondents believed that the wind farm would have had a negative impact in the already built case, while only 5 per cent believed this in the proposed wind farm case.

Dimitropoulos and Kontoleon (2009), however, found contrasting evidence. In a choice model study of wind energy's acceptability on two comparably sized Greek islands, Skyros and Naxos, they found a "considerable divergence" in the perceived threat to tourism between the residents of each island: residents of Naxos denied the risk, while Skyros residents considered it likely (Dimitropoulos and Kontoleon, 2009, p. 1848). This difference can perhaps be explained by the differences in the proposed developments on the islands. Skyros was facing the prospect of receiving ten large wind farms for a total capacity of more than 300 megawatts (MW) on a mountainous area that has some degree of natural protection, and its residents were mostly aware of these plans. The difference of opinions between the two islands suggests that the size of the proposed development and the area's natural value shape residents' beliefs about the proposed wind farm's effects on tourism.

Firestone and Kempton (2007) also asked about the proposed wind farm's impact on tourism when they surveyed the Cape Cod population regarding the proposed offshore wind farm development, Cape Wind, which would have been placed about 5 miles away from the coast. The Cape Cod area has a local economy based on tourism and fishing (Firestone and Kempton, 2007); therefore, the local population would arguably have been sensitive to a threat to tourism. Their survey found that equal proportions (42 per cent) of respondents thought that tourism would be negatively affected or not affected at all, while 8 per cent believed that the project would improve local tourism. When asked about the three main factors that led them to oppose the development, 15 per cent of opponents included tourism; however, most respondents cited potential impacts on marine life/the environment (65 per cent), aesthetics (51 per cent) and fishing/boating safety (50 per cent) (Firestone and Kempton, 2007).

However strange it might appear to some, there is evidence that wind farms actually support local tourism in some cases. Toke (2005b) discusses the case of Swaffham, in eastern England, where the first tall wind turbine installed included a viewing platform. The turbine and the local ecology centre became a tourist attraction, prompting the local parish council to ask the developer for more turbines. A survey detailed in Lothian (2008), which presented Southern Australians with 160 pictures of landscapes with wind farms, corroborates the Swaffham case. Lothian found that for landscapes with low scenic value, wind turbines actually *increased* their perceived value.

Further evidence implying a positive effect on tourism was gathered by Dalton et al. (2008), who surveyed 280 guests at four Australian hotels. Respondents were asked to express a positive or negative opinion (acceptance or rejection) concerning a set of pictures, including one which depicted onshore coastal wind farms near "high-density tourist accommodation" (Dalton et al., 2008, p. 2182). Nearly 68 per cent of respondents expressed a positive opinion of this picture.

More research has focused on the real effects of wind farms on the local tourism industry, rather than their perceived future impacts. Research conducted for the Scottish Government and led by Glasgow Caledonian University using a case study approach to assess the impact on tourism of future wind energy developments concluded: "even using a worst-case scenario the impact of current applications would be very small and for three of the four case study areas, would hardly be noticed" (Glasgow Caledonian University, Moffat Centre and Cogentsi, 2008, p. 16). The Scottish case had been researched previously by MORI Scotland (2002) in a study which investigated the attitudes of tourists in Scotland towards landscapes incorporating wind farms. MORI interviewed 307 tourists in five locations (Tarbet, Inverary, Oban, Campbeltown and Lochgilphead) in September 2002. When tourists were asked whether the presence of wind farms had a positive or negative effect on the landscape, a little over two in five (43 per cent) maintained that it had a positive effect, while a similar proportion felt it was equally positive and negative. Less than one in ten (8 per cent) felt that it had a negative effect (MORI Scotland, 2002, p. 3). About 80 per cent maintained that if a wind farm in the area of Argyll were open to the public and had a visitor centre, they would be interested in visiting the wind farm on another trip.

In conclusion, there is mixed evidence concerning wind farms' perceived effects on tourism. Dimitropoulos and Kontoleon's (2009) study, however, suggests that the size of the development influences perceptions of its anticipated impact.

Local jobs

Wind farm developers and advocates routinely cite local jobs as a benefit deriving from wind farm developments.

Toke (2005b) refers to the case of Lowestoft, in which an engineering firm required planning permission to erect a large wind turbine that would have served as a prototype for future manufacturing. This move attracted a strong consensus in the local community, who viewed the development as a chance to improve prospects on the local job market. Clearly, this could be considered a peculiar case of a likely boon in local jobs.

In their study of the proposed Cape Cod offshore wind farm, Firestone and Kempton (2007) found that 37 per cent of local residents believed the wind farm would have a positive impact on job creation, 28 per cent thought it would have no impact and 27 per cent were unsure, while only 8 per cent believed it would have had a negative impact. Despite this, only 18 per cent of wind farm supporters included "job creation" as one of the three main reasons for their support. Firestone *et al.* (2009) repeated the same study in Delaware with both a state-wide sample and an ocean-area sample. They found that a higher proportion of respondents compared to those in Cape Cod believed the wind farms would have a positive impact on job creation: 48 per cent for the "ocean area" sample and 71 per cent for the state-wide sample. However, only a minority of respondents (25 per cent and 28 per cent, respectively) chose job creation as one of the three major reasons for supporting the project. Furthermore, the researchers asked opponents of the project if they would have been more supportive if they knew that the project would create new jobs. For 44 per cent of opponents, it did not make any difference; 22 per cent said that they would support it "somewhat more"; 17 per cent responded "just a little more"; and another 17 per cent "much more". Compared with other benefits that—if certain—would switch opponents to supporters at the "much more" level, "create new jobs" is in seventh place, the first three being "receive the generated electricity" (33 per cent), "help the local fishing industry" (32 per cent) and "improve air quality on the Cape" (27 per cent).

Jones and Eiser (2009) conducted a hierarchical regression analysis of their survey results, concluding that "employment opportunity" was not a statistically significant economic benefit capable of influencing the support of residents living in the areas adjacent to the proposed wind farms; however, it was significant for residents in the comparison areas located farther away from the sites. The regression analysis demonstrated that job creation had a limited effect on support or opposition, with a standardised beta coefficient of 0.09. On the other hand, the coefficients of 0.15 for "general economic benefit" and 0.50 for "general attitude to wind", the other two variables included in the regression analysis, were statistically significant.

To evaluate the real effect of wind farms on local employment, Munday *et al.* (2011) researched the economic rural development opportunities generated by wind farms in rural Wales. The authors noted that Welsh job gains would be quite modest—perhaps fewer than "150 direct jobs across Wales as a whole" (Munday *et al.*, 2011, p. 6)—especially considering the significant expected investments in large wind farms in the Strategic Search Areas.[2] Their estimate was motivated by the nature of the wind turbine business, which relies on the manufacturer for maintenance, and of the technology used, which requires little onsite maintenance.

Munday *et al.*'s conservative estimate is confirmed by an earlier report by the Centre for Sustainable Energy (CSE, 2005, pp. 12–13), in which the authors note:

> Historic levels of development have been too low and too unpredictable to secure the wider available economic benefits of wind power development in terms of manufacturing and servicing jobs—so there isn't the "it's good for Britain/the region" economic argument.

This situation has partially changed in the last decade, which has seen wind power installed capacity increase substantially in the UK on a year-to-year basis, albeit still at a slower pace than other countries leading the European expansion, notably Germany and Spain (Wind Europe, 2019).

In conclusion, the belief that job creation is a major benefit of wind farm developments does not appear to be widespread; even when respondents recognise it, it does not appear to have a significant influence on their level of support.

Place attachment

Bonaiuto *et al.* (2002, p. 636) define "place attachment" as "the affective relation or the emotional bonds that people have with places where they live". In reality, as Hidalgo and Hernández (2001) point out, there is little agreement within the social sciences regarding an exact definition of place attachment. Hidalgo and Hernández build on a strand of research which points to a both a physical and a social attachment as aspects of the concept of place attachment, a position accepted by research (Devine-Wright, 2009) on the concept and its implications for the social acceptability of wind farms.

Devine-Wright (2009) proposes a framework for understanding the role of place attachment in wind farm opposition. He argues that residents would likely go through a psychological process that can be summarised using the following phases: (i) becoming aware, (ii) interpreting, (iii) evaluating, (iv) coping and (v) acting. Residents would learn about the project through media and discussions with people they trust; would gather more information about it through available sources; and would make sense of the information subjectively. The evaluation phase could lead to alternative subjective stances related to beliefs that the proposed project was either place enhancing or place disruptive. If the project is considered disruptive, coping strategies could be engaged, including active forms of resistance such as joining opposition groups and/or campaigning. However, Devine-Wright considers the outcome to be dependent upon other variables, which are based on research on social acceptability that has postulated variables to explain resistance, particularly political efficacy (Wolsink, 2000) and the presence of cohesive social networks (McLaren Loring, 2007).

Empirical research has looked at how place attachment relates to attitudes about facility siting. For example, Vorkinn and Riese (2001) investigated the role of place attachment in explaining opposition to a large hydro-power station in Norway. They conducted a regression analysis of questionnaire responses using the demographic variables of gender, age, household income and place attachment. Ultimately, they found that place attachment had the highest standardised coefficient (-0.185, $p<0.005$) and explained the highest proportion of the variance, accounting for about 17 per cent of the attitudes concerning the potential development of the hydro-power station.

Devine-Wright and Howes (2010) empirically tested Devine-Wright's (2009) theoretical framework by using in-depth interviews, focus groups and questionnaires to investigate the role of place attachment and trust in shaping social acceptance of Gwynt y Morin, a proposed offshore wind farm in North Wales. The then proposed wind farm (which has since been granted planning permission) consisted of 200 wind turbines for a total installed capacity of 750 MW. They surveyed residents of Llandudno, a seaside resort appreciated for the scenic beauty of its coast and reliant on local tourism, and Colwyn Bay, a former seaside resort with a declining economy. In responding to the questionnaire, residents of Colwyn Bay used negative words or phrases to describe their town: the most frequently mentioned thematic category associated with the town was "being run down". A modest significant correlation (0.22, $p<0.01$) between place attachment and opposition behaviour was found for Llandudno residents, but no correlation was found between the two for Colwyn Bay residents. Bivariate correlations with place attachment were also tested with regard to several other variables. In the case of Llandudno, place attachment was significantly correlated to the statements "create an eye sore", "fence in the bay" and "industrialise the area", while for Colwyn Bay, the only significant correlation was with the statement "help meet national policy targets". Not surprisingly, for persons who reported a high level of trust in the opposition group, there was a negative correlation between place attachment and project acceptance.

Devine-Wright and Howes (2010) conclude that place attachment per se does not inevitably lead to opposition which instead depends on how people interpret change; such interpretation can be influenced by the social context and moderated by how much people trust key organisations. A very high proportion of the survey respondents were older people (the average age of respondents was 61) and, particularly in the case of Llandudno, respondents were slightly older and had spent less time living in the area. It may be that many respondents appreciated the place that they had chosen for their retirement and were therefore averse to the development. Furthermore, they might have wanted to protect the value of their investments if they had bought a property in the area. Conversely, residents of Colwyn Bay might have considered the wind farm an opportunity to generate funding for local regeneration projects, as suggested by an interviewee cited by Devine-Wright and Howes (2010, p. 276). There might also be a correlation between level of affluence (a factor distinguishing the two towns) and level of opposition, as will be detailed later in this chapter.

More recent research has delivered contrasting results with some studies (Devine-Wright, 2013) finding support for the influence of place attachment on local acceptance of energy infrastructure, while others (Devine-Wright and Wiersma, 2020) have failed to find a statistically significant relationship between place attachment and acceptance. Thus, place attachment might have a role in influencing support or opposition to developments, but previous studies have not made fully clear to what extent, and for which type of respondents, it is influential.

Environmental attitudes

Environmental attitudes are clearly significant in determining support for renewable energy, as these attitudes make individuals more aware of the dangers of climate change and other polluting effects of fossil fuels. However, it has been pointed out (Warren *et al.*, 2005) that some people might be more disposed to protecting their local environment over the global environment. In expressing underlying pro-environmental attitudes, such individuals tend to oppose wind developments.

Swofford and Slattery's (2010) survey of residents living close to the Wolf Ridge wind farm in Texas found that respondents displayed high general environmental concern: roughly 93 per cent agreed with the statement "protecting the environment is important to me", while only about 58 per cent agreed with the statement "I am concerned about climate change". About 60 per cent of respondents evinced positive attitudes regarding wind energy. Of those respondents concerned about climate change, 63 per cent expressed support for wind power, while 17 per cent were against it. Of those respondents unconcerned about climate change, 46 per cent expressed support for wind power, while 27 per cent were against it. This study, therefore, suggests that climate change concerns may influence support for wind power, albeit only to a limited extent.

In her study of residents living close to four Scottish wind farms, Dudleston (2000) created a scale of environmental concern based on six questions: use of recycling facilities; membership of organisations committed to environmental protection; concern with damage to the countryside; concern about the loss of wildlife; concern about global warming; and concern about the depletion of natural resources such oil, gas and coal. Positive answers received one point, thereby placing each respondent on a scale from six (high environmental concern) to zero (low environmental concern). Dudleston (2000) found that respondents who displayed low environmental concern were often indifferent to wind power, while respondents with high environmental concern generally expressed opposition to wind farms. One question regarded the level of concern the respondent would have if another wind farm were proposed in the area. The relative majority of respondents who were not at all concerned about it scored low in the environmental concern scale, while the relative majority of very concerned and fairly concerned respondents scored high on the environmental concern scale.

Pro-environmental attitudes do not necessarily indicate support for wind farms, although they can. There may be a divide between people who view wind energy favourably as a good source of clean energy that needs to be developed in order to tackle climate change, and people who are very concerned about preserving the features of their local environment.

Perceived non-local costs and benefits

Factors of non-local relevance include those related to the collective benefits and costs of wind energy at a societal, global or national level. The most prominent among them are perceptions regarding wind energy's role in reducing pollution, particularly CO_2 emissions; the impact of wind energy growth on national energy security; the economic costs of energy produced by wind power; and wind power's reliability compared with other energy sources.

Decarbonising the energy supply

Reducing CO_2 emissions is wind electricity's most touted positive impact at the global level. However, the production of electricity from non-fossil fuel sources can also bring other important positive environmental impacts if we consider that electricity through fossil fuel production inevitably releases several pollutants which have a significant environmental impact (Williams, 1993) and consequently an impact on human health.

Wolsink (2007b) investigated the weight of wind energy's perception as a clean energy source in determining the acceptability of proposed wind farms. In his LISREL analysis of 725 responses to a survey investigating the motives for local wind farm acceptability, Wolsnik found that the variable "wind as clean energy source" was not directly correlated to the variable "resistance to local wind developments", but was correlated to the "wind power attitude" variable. Moreover, the variable "landscape/visual" was also correlated with "wind power attitude" with a stronger correlation index than that of "wind as clean energy source", implying that aesthetics play a greater role in determining "wind power attitude" and ultimately resistance to a local development.

Offshore wind farms, like their onshore counterparts, often receive considerable opposition. Firestone and Kempton (2007) found, surprisingly, that residents scarcely mentioned "global warming/climate change stability" among their top three reasons for opposing or supporting the proposed offshore wind farm at Cape Cod: just 4 per cent of supporters and 4 per cent of opponents had considered this issue as one of the top three informing their judgements. The situation changed when the survey was extended to Delaware residents (Firestone *et al.*, 2009): 21 per cent of supporters residing on the coast indicated that "global warming/climate change stability" was one of the top three factors influencing their opinion, while 12 per cent of supporters in the state-wide sample deemed it influential. Importantly, among both coastal and inland residents state-wide, no opponents cited wind power's impact on climate change as one of the top three factors contributing to their opposition.

Even in an extended sample, Firestone *et al.*'s (2009) study shows that, at least for Cape Cod residents, factors other than climate change mitigation, including "marine life/environmental impacts" (48 per cent), "electricity rates" (47 per cent) and "foreign oil dependence" (37 per cent), were more influential in determining their support for the proposed wind farm. Among residents of coastal Delaware, "electricity rates" (56 per cent of respondents), "air quality" (46 per cent) and "marine life/environmental impacts" (29 per cent) were the most-cited reasons for support. Among the state-wide inland sample, "electricity rates" (62 per cent), "marine life/environmental impacts" (57 per cent) and "air quality" (39 per cent) were most cited. Thus, concerns that more closely affect respondents, in terms of their geographical context, are more important in determining support than climate change mitigation. Nonetheless, it is also true that other environmental issues are one of the first three most frequently cited motives for support, particularly "marine life/environmental impacts," which was mentioned in all three samples, and "air quality," which appeared in both the coastal Delaware and state-wide samples. Interestingly, "electricity rates" are very highly rated within the three samples, and they are actually the most reported motive for support for the two non-local samples (coastal Delaware and state-wide). This misconception that wind energy delivers cheaper electricity has been suggested as one reason for wind power's popularity in the USA (Klick and Smith, 2010).

In their study of a national sample of 610 adult US citizens, Klick and Smith (2010) asked respondents several questions regarding their knowledge of wind energy and their agreement with wind farm's positive or negative impacts. They found that wind energy's positive aspects attracted more agreement than its negative aspects. Around 80 per cent of respondents rated as "very important" the statements "emit no pollution" and "emit no greenhouse gases", while the highest percentage of agreement with a negative statement was attributed to "more expensive", which 31 per cent of respondents rated as "very important" and 46 per cent as "somewhat important". In one of the two regression analyses conducted, the two leading factors which determined support were "reduce imported energy" (0.26, p<0.01) and "emit no pollution" (0.20, p<0.05), while the two factors which had the strongest negative impact on support were "lower property values" (–0.16, p<0.01) and "more expensive" (–0.16, p<0.01). Visual impact seemed to have a non-statistically significant impact of modest magnitude. Importantly, however, this was a national study that inquired about wind energy in general. Research demonstrates that resistance towards wind energy is stronger when residents are surveyed about specific, locally proposed projects (Haggett, 2004; Wolsink, 2007b).

Swofford and Slattery's (2010) survey of residents within a 20 km radius of Cooke County, Texas Wolf Ridge wind farm, found that about 60 per cent expressed support for wind energy, while about 18 per cent had a negative attitude. Overall, 58.4 per cent of respondents agreed with the statement "I am concerned about global climate change": 63.3 per cent of respondents who supported wind agreed, while only 17.4 per cent of those who expressed a negative attitude towards wind agreed with this statement. For those respondents not concerned about climate change, 46 per cent still supported wind energy, while 27 per cent did not. Only 34.5 per cent agreed to support renewable energy, even if it cost more than energy from fossil fuels. Forty-seven per cent considered wind farms "an unattractive feature of the landscape", while 20.5 per cent disagreed and 22.3 per cent were neutral. Texas residents' support and opposition were motivated by wind energy's contribution to the mitigation of climate change; however, their support fell if it was stated that wind and other renewable sources were more expensive than fossil fuels in producing electricity. Furthermore, the visual impact of a wind farm was a polarising issue for respondents, with a large number saying that it was an unattractive landscape feature.

In sum, research has shown that decarbonising energy supply is one factor that influences support for wind farms, but it appears that other concerns related to wind farms' local environmental impact, or economic considerations about their development, are more important in shaping support or opposition.

Improving air quality

Wind farms also improve air quality because they do not emit particulate matter and other pollutants, unlike power plants, which produce electricity from the combustion of fossil fuels. In the future, a large deployment of renewables, combined with mass scale development and commercialisation of electric cars, could lead to the use of fewer combustion engines for transport and, as a result, better air quality and improved public health. At the moment, however, this possibility is far from being realised, and the marginal improvement in air quality that would result from the construction of a single wind farm would likely go undetected by anyone living close to the wind farm unless, for example, the wind farm was developed on land earlier occupied by a combustion power plant.

This advantages of wind power as well as other societal advantages of wind energy have been discussed in the literature (Toke, 2002; Sustainable Development Commission, 2005), but this subject has not been investigated very much.

In their survey regarding the Cape Cod offshore wind farm proposal and a similar hypothetical development in Delaware, Firestone *et al.* (2009) found that 23 per cent of Cape Cod residents, 46 per cent of coastal Delaware residents and 39 per cent of Delaware residents state-wide considered "air quality" to be one of the three most important factors influencing their decision to support a wind farm. For wind farm supporters in the Cape Cod area, "air quality" was the fifth most cited factor, while among Delaware coastal-area residents and state-wide respondents, it was the second and third most cited factor, respectively. "Electricity rates" rather than "air quality" were consistently cited as one of the three most important factors across all samples. In the Cape Cod and Delaware state-wide samples, "marine life/environmental impacts" was nominated more than "air quality". In all three cases, "global warming/climate stability" was cited as one of the three most important factors, considerably less than was "air quality", which received 4 per cent of positive responses for Cape Cod, 21 per cent for the Delaware coastal area and 12 per cent for Delaware state-wide.

Hence, while air quality could influence support for wind energy, like "decarbonising energy supply" it is probably less important than local environmental or economic factors, but more important than concerns about climate change.

Costs of electricity production

Several studies have considered the cost of electricity produced by wind energy (Firestone and Kempton, 2007; Firestone *et al.*, 2009; Jones and Eiser, 2009; Klick and Smith, 2010; Swofford and Slattery, 2010).

Cheaper electricity rates consistently motivated support for wind in the three samples (Cape Cod area, Delaware ocean area and Delaware state-wide) surveyed by Firestone *et al.* (2009) and discussed above: electricity rates were the two most reported reasons for wind farm support among all samples.

Jones and Eiser (2009) also found that "cheaper electricity" influences support for wind power. They ran a hierarchical regression analysis of hypothetical benefits deriving from a proposed wind farm while controlling for "general attitude" towards wind energy. They found that for residents living in areas adjacent to the proposed sites, "cheaper electricity" was a statistically significant factor leading to support (0.13, p<0.010); however, this factor had a lower standardised coefficient than "general attitude" towards wind (0.32), "general economic benefit" (0.30) and the "opportunity to invest" (0.13).

Similarly, Swofford and Slattery's (2010) survey of residents close to Texas' Wolf Ridge wind farm found that while about 60 per cent of respondents held favourable views towards wind and nearly 84 per cent agreed that "we should use more renewable energy to fulfil US energy demands", only 34.5 per cent agreed with the statement "I am willing to support renewable energy even if it costs more than energy from fossil fuels". This finding suggests that support is conditional on the cost of electricity to consumers.

Finally, using a regression model to explain support for wind, Klick and Smith (2010) found a significant negative correlation (–0.16, p<0.01) between the belief that wind energy is "more expensive" and support. It is important to note that this correlation was found only for the regression analysis of the support variable related to the question of support/opposition presented at the end of the questionnaire; data collected from the same question presented at the beginning of the questionnaire did not present a significant correlation with the "more expensive" variable. It is likely that the questionnaire induced reflection on the pros and cons of wind energy.

Thus, the cost of electricity produced by wind energy appears to be an issue many individuals consider when determining support for or opposition to a wind farm, though it is not clear how much it motivates their support or opposition.

Reliability of wind energy supply

Part of what motivates opposition to wind energy is the perception of its unreliability (Country Guardian, 2000). The UK Sustainable Development Commission (2005) played down the significance of the issue, highlighting the technical feasibility of increasing the share of wind energy generation in the power supply, and explaining that increased costs due to the output variability of wind generation would be limited unless the share of wind energy supply were to become very large.

Devine-Wright and Devine-Wright (2006) investigated how organisations and campaign groups communicated the issue of intermittency to the public. They found that the vocabulary used to communicate intermittency depended upon whether the communicators supported or opposed wind energy: opponents usually highlight wind energy's "unpredictability" and "uncontrollability", while supporters talk instead of its "variability" and "fluctuation".

In their comparative study of community wind farms in Scotland, Warren and McFadyen (2010) asked both of their samples in Kintyre and Gigha (in the Argyll and Bute area) to identify their two greatest concerns about wind power. The category that received the largest response was "no concern" (48 per cent in Kintyre, 32 per cent in Gigha) followed by, in order of decreasing importance, intermittency of production, visual impact and bird strikes on turbines and habitat disruption (the authors did not provide percentages with regard to these variables).

The evidence concerning wind energy's reliability and its influence on support or opposition is limited. However, it is important to keep concerns about reliability in mind, since opponents frequently cite wind energy's unpredictability, and respondents may have heard such arguments.

Energy independence and security

According to Awerbuch (2006, p. 693), energy security is concerned with the "threat of abrupt supply disruptions", although he proposes to extend the meaning to the risk created by fossil fuel price volatility. He argues that adding increasing shares of wind, geothermal energy and other renewables to the energy generation portfolios of European Union and American economies would increase these countries energy security both in terms of energy supply and generating costs.

Eltham *et al.* (2008) found that residents of St Newlyn East in Wales, about 2250 metres from the Carland Cross wind farm (a cluster with 6 MW of total capacity and 15 turbines built in 1992), considered "energy security" the most positive of the set of self-identified positive impacts that they reported. Cited by 41 per cent of respondents, energy security ranked ahead of "visual attraction" (40 per cent) and "green power" (22 per cent). Just 6 per cent of respondents had considered energy security a positive impact during the planning phase in 1991 (Eltham *et al.*, 2008), demonstrating a vast change in the perception of the importance of this issue over the years.

Energy security might have a secondary impact on local opposition or support, mainly because a single wind farm would not contribute in a large way to alleviating this problem, yet it is a relevant argument influencing the debate.

Environmental factors of local relevance

Environmental factors of local relevance, particularly visual impact on the local landscape, audible noise and non-audible, low-frequency noise, are frequently cited in academic and non-academic publications as sources of local opposition. These factors, examined below, are considered attitudinal because they are relevant, for this research, as anticipated effects of a future wind farm rather than measurable physical phenomena that impact residents' lives.

Visual impact

Local visual impact is crucial in determining opposition to a proposed wind farm (Warren *et al.*, 2005; Wolsink, 2007a). According to Devine-Wright (2005a, p. 127), "Research attempting to identify possible reasons for public opposition to wind farms has noted visual impacts and noise as the most frequently reported problems."

In a survey of residents close to the Dun Law wind farm in the Scottish Borders, a project of 26 turbines for a total of 17.2 MW of installed capacity that was under construction at the time of the survey, Warren *et al.* (2005) found that residents considered visual impact as both the most significant positive and negative impact (34 per cent believed the turbines were an "attractive feature in the landscape", while 44 per cent indicated that turbines were an "unattractive feature in the landscape") of the wind farm.

Wolsink (2007b) reports the findings of a study in which Dutch respondents were asked to rank the significance of several issues concerning wind power on a 7-point Likert scale. The regression analysis of the data set indicates that visual impact has a far larger impact (0.48 p<0.001) on "wind power attitude" than the other variables examined ("decrease environmental issues" 0.19 p<0.001, "annoyance" 0.17 p<0.01, "electricity sector" 0.07 p<0.05). When the same respondents were asked about their reason for opposing a proposed local wind farm, Wolsink's analysis showed how "landscape visual impact" had the largest impact on both the variables "wind power attitude" (0.74) and "resistance to local wind developments" (–0.20), within a group of variables which also included "wind as clean energy source", "fairness/equity/NIMBY",[3] "personal political efficacy" and "annoyance".

In another study, Meyerhoff *et al.* (2010) investigated public perceptions of onshore wind farm developments, providing different options for a site's size and distance from residential areas. The authors used a choice model method to conduct 708 phone interviews, roughly equally divided between the regions of Westsachsen and Nordhessen in Germany. They found that the regions' perceived environmental quality did not influence perceptions, nor did respondents' proximity to the wind farms or the frequency with which they encountered turbines in the landscape. However, the authors found that, on average, respondents would prefer turbines to be further away from residential areas and negatively valued their impact on biodiversity. Unexpectedly, the height of the turbines did not influence respondents' choices; this finding contrasts with the expectation that respondents would have preferred smaller turbines because of the reduced impact on the landscape.

Devine-Wright (2005a, p. 127) mentions a number of studies that have highlighted the relationship between size of the wind farm (i.e. the number of turbines) and support; he notes that there is a negative linear relationship between the size of the wind farm and support. In fact, a study conducted in Ireland by Sustainable Energy Ireland (SEI, 2003) showed that there was a preference for smaller clusters of turbines, rather than large ones: interestingly, smaller clusters of taller turbines were preferred over larger clusters of shorter turbines.

As noted earlier, there is some evidence of positive appraisals of a wind farm's impact on the landscape (Warren *et al.*, 2005). This seems to relate to subjects' differing perceptions of the landscape in which the wind farms are situated. Research conducted by Lothian (2008) showed that wind farms are believed to enhance landscapes of perceived low scenic quality and detract from landscapes of perceived high scenic quality. Nevertheless, this finding is not corroborated by SEI's (2003) study. Utilising the same procedure as Lothian's study, researchers showed Irish respondents pictures of different types of landscapes which were first presented without turbines and rated in terms of amenity. Wind farms did not negatively affect respondents' perceptions of the landscapes considered to be the most beautiful. For example, respondents rated the wind farm's scenic impact on the "costal area" as "very positive" (15 per cent), "fairly positive" (29 per cent), "neutral" (29 per cent), "fairly negative"

(19 per cent) and "very negative" (7 per cent). Similarly, wind farms on less beautiful landscapes showed a similar response pattern: in the case of the "urban/industrial area", considered the least beautiful of all, respondents believed that the wind farm's scenic impact was "very positive" (13 per cent), "fairly positive" (32 per cent), "neutral" (33 per cent), "fairly negative" (15 per cent) and "very negative" (7 per cent).

In conclusion, while visual impact is a significant element in influencing respondents' views about wind farms, there is no consensus regarding how a wind farm affects the landscape. While some people believe that a wind farm enhances the value of a landscape, others believe that it detracts from it. Yet there is some evidence that people prefer smaller clusters of turbines over large clusters of turbines.

Noise

Noise is one of the most frequently reported reasons for opposition to wind energy (Devine-Wright, 2005a). However, few studies have investigated the significance of noise in determining opposition to wind farms before they are constructed. In one such study, Wolsink (2007b) asked 725 Dutch respondents about their intentions and motivations to oppose a local wind farm. The researchers performed LISREL analysis including the following factors: "landscape/visual", "wind power attitude", "wind as clean energy source", "fairness/equity/nimby (not in my backyard)", "personal political efficacy" and "annoyance", which includes the factors of noise, light and flicker. "Annoyance" was found to be the third most important factor influencing "resistance to local wind developments" after the variables "landscape/visual" and "wind power attitude", which was itself substantially determined by "landscape/visual".

In another study, Eltham *et al.* (2008) investigated changes in attitudes towards the Carland Cross wind farm development in Cornwall, England between the pre-construction and post-construction phases. Respondents were asked about their attitudes towards the development pre-construction and post-construction and to rank each element on a 5-point Likert scale. The results indicated that the most feared negative impact before construction was noise, which was perceived as a probable negative impact by 23 per cent of respondents but decreased to 15 per cent when respondents were asked about their post-construction perception (a statistically reliable change). The second most feared negative impact was visual intrusion at a level of 19 per cent pre-construction and 15 per cent post-construction. Thus, this research found that residents had a different ranking of the importance of perceived expected negative impacts; earlier studies had indicated that visual impact was the most important anticipated negative outcome (Warren *et al.*, 2005; Wolsink, 2007b).

Based on a sample of residents living around two already built wind farms in southwest Ireland, Warren *et al.* (2005) found that the most feared expected negative impact was visual impact (89 per cent) followed by noise (59 per cent). Eltham *et al.* (2008) hypothesise that the difference in their finding is owing to the fact that the wind farms researched by Warren *et al.* were built later, using larger turbines. Furthermore, they offer, as another reason, the fact that respondents in the Warren *et al.* (2005) study would have had the chance to experience first-hand, or through the media, the visual impact of wind farms, while the villagers of St Newlyn East would not, because their wind farm was built in 1992 at a time when the UK public knew little about wind power.

The findings of Eltham *et al.*, however, are not isolated. Exeter Enterprises (1994), which surveyed 435 residents within a 2 km radius of the wind farm site of Delabole (a cluster of ten turbines) in Cornwall before (1990) and after (1992) construction found that noise was the most important expected negative impact. However, this concern fell sharply (from 86 per cent down to 20 per cent after construction), while visual impact was less of a concern (about 50 per cent pre-construction, and still considered a problem by 25 per cent after construction). Still, the majority of the surveys reviewed indicate that noise was an anticipated negative outcome of lesser importance than visual impact.

Beyond the earlier cited studies (Warren *et al.*, 2005; Wolsink, 2007b), others report a higher concern with expected visual impact than with noise. For example, Dudleston (2000) surveyed 430 residents living within 20 km of four operational wind farms in Scotland. She found that noise before construction was recollected as a concern by 12 per cent of respondents, second in importance after visual impact (27 per cent), while the percentage of respondents who disliked the noise impact of the wind farm was only 2 per cent post-construction. Similarly, Braunholtz and McWhannel (2003), who surveyed 1810 residents living within 20 km of ten large Scottish wind farms, found that noise was anticipated to be a problem by 12 per cent of respondents, but only 2 per cent considered this to be a problem after construction.

Eltham *et al.* (2008, p. 30) point out that the proportion of the surveyed population who anticipate that noise will be a negative impact varies greatly from 12 per cent (Dudleston, 2000) to 59 per cent (Warren *et al.*, 2005) to 86 per cent (Young, 1993). Obviously, these proportions are far higher in the studies of Warren *et al.* (2005) and Exeter Enterprises (1994) (59 per cent and 86 per cent respectively) than those in the other studies referenced. The explanation is, perhaps, that in the cases of the two studies cited above respondents were surveyed *before* the construction of the wind farm, i.e. they were not interviewed post-construction and asked to recall their opinion before the wind farm was built. This might have generated answers which were not influenced by the process of recalling old memories, which, particularly if the attitude towards noise had changed (i.e. noise was not considered a problem after construction), might have produced in respondents the desire to represent their old attitudes as consistent with their new ones, thus minimising their previous negative expectations about the impact of noise in the pre-construction phase.

The decreased perception of the negative impact of noise from a pre-construction to a post-construction phase is consistent in all the studies reviewed that compared pre-construction to post-construction attitudes (Young, 1993;

Dudleston, 2000; Braunholtz and McWhannel, 2003; Eltham et al., 2008), a change of attitude noted and discussed in Warren *et al.* (2005).

One specific type of noise that is of concern is low-frequency noise which can be audible or inaudible. It has received consideration in research as a type of environmental impact of wind farms. Low-frequency noise emitted from wind farms could theoretically produce negative health consequences for those exposed to it. The use of the low-frequency noise argument by anti-wind farm activists can be traced in the materials they have produced (see, e.g., Country Guardian, 2000). However, *surveys* that specifically studied the influence of this expected negative impact on the social acceptability of proposed wind farms were not traced.

Research on the low-frequency noise produced by wind turbines seems to exclude any risk to human health (Bellhouse, 2004; Roberts and Roberts, 2009). Wind turbine noise is considered more of a public annoyance than a health risk (Pedersen and Persson-Waye, 2004; Pedersen *et al.*, 2007; Pedersen and Persson-Waye, 2007; Pedersen and Persson-Waye, 2008) while Janssen *et al.* (2009) found that annoyance caused by wind farm noise is related to sound pressure levels, the site's environmental features and residents' attitudinal disposition towards the wind farm.

Pedersen and Persson-Waye (2004) found a significant correlation between attitude towards the visual impact of the wind farm and "noise annoyance" in their analysis of a 2000 Swedish survey; their findings were confirmed by later research conducted by Pedersen and Larsman (2008). This found that "noise annoyance" has the highest positive correlation coefficient with "visual attitude" for residents who can see the wind farm from their home. Similarly, in Janssen *et al.*'s (2009) study, respondents who reported that they could see one or more wind turbines from their home also reported higher levels of annoyance.

Perhaps not surprisingly, subjects who benefit economically from a wind farm appear to perceive them as less of an annoyance. Janssen *et al.* (2009) analysed data from three surveys (one in Holland and two in Sweden) and found that respondents who partly own one or more wind turbines reported less annoyance with the noise.

The studies reviewed make clear that noise is a common concern during the pre-construction phase and is one of the factors most related to a locally proposed wind farm's acceptability. Nevertheless, noise concern is usually second to the development's visual impact and diminishes drastically during the post-construction phase: in other words, many respondents do not consider noise as a real problem after construction.

Personal resources

Personal factors—specifically, knowledge about wind energy, affluence or deprivation, education and proximity to the proposed wind farm site—can also affect respondents views of a wind farm.

Knowledge about wind energy

Knowledge about energy production and renewable energy, particularly wind energy, can influence public attitudes in general as well as those of residents living close to wind farm developments. This might appear obvious: this chapter has already highlighted how beliefs about wind energy, which derive from either correct or incorrect information, can influence acceptability; in particular, beliefs about the economic costs of electricity produced through wind power or the possible negative health effects of wind turbines on the local population.

The larger the number of issues that are misrepresented to the public, the greater the likelihood of a negative (or positive) impact on wind farm acceptability, even at the local level. Klick and Smith (2010) asked four questions regarding knowledge of wind energy. There was a poor correlation between correct responses across the items, indicating that subjects answered some questions correctly and others incorrectly. Furthermore, the authors found that knowledge of wind power's public benefits was lacking. They suggested that a more accurate understanding of the issues might alter public support. While it can be safely assumed that local impact comes into play when the public is surveyed regarding a local wind farm project, it is still reasonable to believe that knowledge about wind energy will, to some extent, influence their attitudes towards the development. In a meta-analysis of psycho-social determinants of pro-environmental behaviour, Bamberg and Möser (2007, p. 22) found that knowledge and awareness are "important indirect determinant[s] of pro-environmental behaviour".

Affluence/deprivation

As has been mentioned, Toke (2005b) notes that different subjects, belonging to differing social classes, might come to perceive wind farms in different ways. On the one hand, farmers may see in wind farms an opportunity, while, on the other, affluent middle-class residents might want to protect landscape views and, ultimately, the economic value of their properties.

In a literature review on the concept of "NIMBYism" (not in my back yard), van der Horst (2007) highlighted the differences between rural areas of higher landscape value, which are often within commuting distance of economically active areas, and rural areas that are more remote and/or have low landscape value due to their industrial heritage. Well-connected areas with high landscape values are likely to attract affluent middle-class residents who wish to retreat into the countryside in pursuit of the "rural idyll", while low landscape value rural areas are more likely to be inhabited by less affluent or deprived residents, who may be used to industrial facilities and might, therefore, accept the presence of wind farms. Conversely, wealthier residents in high-value landscape areas might resist wind farms in their area to protect "their financial and emotional investment" (van der Horst, 2007, p. 2709). van der Horst and Toke (2010) also highlight different classes' varying abilities to lobby against proposed wind farms. They found that areas that are more likely to refuse planning permission have populations with a higher life expectancy, higher voter turnouts and lower crime privation indexes, i.e., lower exposure to crime. If the analysis is extended to the appeal stage, other variables come into play, namely: a lower health deprivation index; a smaller percentage of people aged 16–24; a higher proportion of those who are self-employed; fewer students; a lower proportion of people working in public administration, defence and social security; lower levels of illness and disability; fewer road accidents; and a higher proportion of second homes and holiday homes. It seems, therefore, that areas which are more affluent and less deprived resist proposed wind farms in their areas more often or more efficaciously. This finding is consistent with research on LULU (locally unwanted land use), which finds that deprived areas are more likely to host unwanted facilities (van der Horst and Toke, 2010). Whether these findings are due to residents' higher political efficacy or to higher levels of opposition is debatable.

Firestone and Kempton (2007) found that opponents of wind farms were more likely to have incomes above US$200,000 per year, while residents in the income bracket of US$150,000–199,999 were 20 times more likely to support a wind farm. Those with lower incomes (up to a lower threshold of US$35,000 per year) were also more likely to support a wind farm, although the relationship was not statistically significant in this latter case.

In his review of published research on environmentally significant behaviours, Stern (2000) states that socio-demographic variables such as age, educational attainment, race and income have limited explanatory power on many environmentally significant behaviours, except for those that depend on particular capabilities. In Stern *et al.* (1999), he offers the example of "environmental citizenship", a set of active pro-environmental behaviours (e.g. signing petitions, boycotting companies or products, donating to environmental groups and the like) that are positively associated with income. As Warren *et al.* (2005) indicated, both support and opposition to wind farms can be considered two forms of environmentalism and, regarding opposition, van der Horst and Toke (2010) suggest that this can be stronger and more effective in affluent areas.

Affluence and deprivation, therefore, appear to have an impact on social acceptability; it has not yet been established whether this is triggered by the intent to defend a property investment or simply to defend the amenity of a locale. It is certainly true that less deprived areas are often unspoiled countryside environments that might trigger opposition in and of itself when their integrity is perceived to be under threat.

Education

Given the nature of co-operative wind farms, education can also play a role: the more educated subjects are, the more likely they are to understand the co-operative model and its social and financial implications.

Research has shown that higher education levels correlate to pro-environmental attitudes and behaviours. Barr *et al.* (2005) studied household pro-environmental behaviours and found that the cluster of respondents defined as "committed environmentalists" were middle income and included a higher number of degree holders than "mainstream environmentalists", "occasional environmentalists" and "non-environmentalists". Shen and Saijo (2008) found that respondents with a college degree or above evinced significantly more environmental concern than other respondents.

Proximity of residence and time of survey

Research on wind farms (Krohn and Damborg, 1999; Braunholtz and McWhannel, 2003; Warren *et al.*, 2005; Bishop and Miller, 2007; van der Horst and Toke, 2010) demonstrates that before construction, residents who live closer to the development tend to be more concerned and negative about the project than residents who live further away.

Residents' proximity to a wind farm site influences levels of opposition and support, and this changes over time in relation to the construction of the wind farm, with local residents displaying greater support after construction and less support before construction (Warren *et al.*, 2005; van der Horst, 2007). Devine-Wright (2005b) discusses contrasting studies on the influence of physical proximity on local residents' perception of wind farms, but his review is limited to social research on residents' attitudes once the turbines are in place.

Warren *et al.* (2005) propose what they call an "inverse NIMBY" syndrome, which holds that after construction individuals living closer to a wind farm display more support for it than residents living further away. The authors surveyed residents living close to four wind farm sites in southwest Ireland and southeast Scotland; three had a wind farm already constructed at the time of the survey, while one of them did not. They found that residents close to the Black Hill proposed wind farm site displayed more opposition than residents living close to the constructed Dun Law wind farm. Furthermore, in the case of Black Hill, residents living closer to the proposed site displayed greater opposition (33 per cent), than residents living further away (3 per cent). In the case of Dun Law, the inverse was true, with residents living 0–5 km from the wind farm area displaying less opposition (6 per cent) than residents who lived 5–10 km from the wind farm (10 per cent). An explanation of this difference is that residents living close to the site gained direct experience of the wind farm and changed their attitudes after construction, while the same was not true for those who live further away (van der Horst, 2007).

Dudleston's (2000) findings are consistent with the Warren *et al.* (2005) study. The author found residents living in close proximity to four operational Scottish wind farms were more pleased with them: 67 per cent overall said that there was something they liked about their local wind farm, while the proportion rose to 73 per cent in the high proximity zone of 0–5 km from the wind farm. Dudleston also found that the proportion of residents who anticipated problems with the wind farm was far higher (40 per cent) than the proportion of residents who actually reported problems once it had been constructed (9 per cent).

In their survey of residents living close to ten large operational wind farms, Braunholtz and McWhannel (2003) confirmed that residents living closer to a local wind farm hold more positive views of it. When asked what effect the wind farm had on their local area, 45 per cent of residents living within 0–5 km of the wind farm site said they felt "generally positive" or "completely positive" about the wind farm; this percentage declined to 42 per cent in the 5–10 km area and further declined to 17 per cent in the 10–20 km area.

Eltham *et al.* (2008), found that despite the non-significance of the result, a change was registered over time with regard to residents' opinions about a wind farm; the percentages of objection in the sample decreased from 14 per cent to 6 per cent, while those of support increased from 74 per cent to 82 per cent.

Wolsink (2007b), however, attempts to refute the link between vicinity and differing attitudes by presenting a Dutch case—although his arguments do not sound convincing. In fact, in the study he discusses, the sample is composed of members of a national organisation that has the aim of protecting an ecologically important Dutch area. Therefore, the members' attitude towards the presence of turbines in this area is not influenced by their proximity to it, but rather by their membership in the organisation, whose members all hold similar attitudes towards the protection of the area.

The findings of several authors corroborate the "inverse NIMBY" syndrome proposed by Warren *et al.* (2005). This is clearly a personal factor, the importance of which importance cannot be downplayed. It needs to be addressed by policymakers, who will need to guarantee fair treatment of the communities most proximate to any proposed site.

Contextual factors

Contextual factors denote all variables that are factual and related to the context of the development.

Issues of procedural justice and participation in a co-operative and other types of community ownership or community benefits schemes are discussed here. All of these factors relate to real processes that occur before construction and have social, psychological and economic implications. Clearly, the effect of these factors on acceptability is principally an effect on attitudes; therefore, they could also be subsumed under the category of attitudinal factors. The distinction is maintained because they refer to real processes that occur in the context of a community wind farm project, and because they develop during the planning phase; therefore, they present themselves as a reality to residents, rather than as a hypothetical consequence of a wind farm, the real impact of which can only be gauged once it is built.

Procedural justice

Various authors advocate for a process of project development and planning which would engage and inform the community, thereby enhancing a sense of fairness and leading to more favourable attitudes towards a wind farm project. This process is referred to here as "procedural justice".

Gross (2007, p. 2729) states that: "Important elements in procedural justice include rights of participation, access to information, and lack of bias on the part of the decision-maker." She studied the case of Taralga in New South Wales, Australia, interviewing members of the local community about a proposed 69-turbine wind farm development. Interviewees complained of a lack of real consultation and about the paucity of information provided. The interviewees contested the fairness of the consultation process identifying three main issues: "secrecy", "insufficient community discussion" and "inequitable distribution of benefit". Secrecy was related to the complaint of "insufficient community discussion", namely, a lack of broad involvement of the local community. Distinguishable from the previous two, which relate to the process, the third complaint, "inequitable distribution of benefit", instead points to a lack of equity in the outcome of the wind farm consultation process: this was presented as a "foregone conclusion" (Gross, 2007, p. 2733). The author concludes that the principles of procedural justice, which include appropriate participation, ability to be heard, adequate information, being treated with respect and unbiased decision-making, were considered important by interviewees (Gross, 2007, p. 2736).

In their qualitative case study of the municipality of Zeewolde, a municipality and town in the Flevoland province in central Netherlands, Agterbosch *et al.* (2009) concluded that the perceived fairness of the planning process and equitable benefits are important factors driving support or opposition. They state:

> [t]he building of a network of administrative and public support and collaborative arrangements, the ability of voice [*sic*] to be heard by local stakeholders and an adequate dissemination of information are important social conditions that add to a sense of procedural justice.
>
> (Agterbosch *et al.*, 2009, p. 404)

Specific features of a procedural justice approach are sometimes highlighted as significant in facilitating support. Krohn and Damborg (1999) emphasise the importance of communication, dialogue and information in preventing opposition to wind projects. The authors cite a study (Hoepman 1998, cited in Krohn and Damborg, 1999) carried out in Friesland, Holland, in which 85 per cent of respondents indicated that they wanted to be informed of new wind energy developments and 60 per cent thought that informing the community was the duty of the local authorities. Only 13 per cent thought that this was a task for the media.

In conclusion, procedural fairness could be conducive to the wider acceptance of a proposed wind farm, particularly because it would lend a sense of transparency to the whole wind farm project, rather than appearing as something done behind the back of local residents and purely for the profit of the developer.

To be invoked, procedural fairness seems to require some basic elements: widely distributed information, engagement of residents and a fair handling of the participation process to prevent an undesired outcome.

Trust towards the proponents

Trust towards the proponents and the institutional actors involved in a wind farm development has been indicated by several authors to be a variable which influences levels of support (Wolsink, 2007a; Devine-Wright, 2008; Toke *et al.*, 2008; Agterbosch *et al.*, 2009; Jones and Eiser, 2009; Aitken, 2010; Devine-Wright and Howes, 2010).

Devine-Wright and Howes (2010) found a significant correlation between trust towards the developer and opposition to the proposed wind farm (−0.191, $p<0.003$). Agterbosch *et al.* (2009) pointed to trust and procedural fairness as elements which, along with a community stake in the project, could increase consent. Similarly, Jones and Eiser (2009) carried out a hierarchical regression analysis that had the attitude towards a proposed development in Sheffield as its dependent variable and which controlled for the "general attitude" to wind variable. The variable "trust" was calculated using six items that asked respondents about their trust towards Sheffield City Council that had proposed the wind farm developments. Respondents were asked whether they trusted Sheffield City Council to: (i) seek local opinion; (ii) take local opinion into account; (iii) keep residents views at heart; (iv) keep locals informed; (v) tell the truth about any risks; and (vi) act fairly when choosing a final site (Jones and Eiser, 2009, p. 4609). The authors found that "trust" was a predictor of support for the target group (residents living in areas adjacent to the proposed sites) with a standardised coefficient of 0.22 ($p<0.001$) which compared with 0.55 ($p<0.001$) for "general attitude" to wind.

In conclusion, "trust" is a factor that influences support, but it must be considered in the context of other factors at stake. Further research is essential in order to clarify the extent of its impact.

Types of community benefits

Another contextual (and economic) factor that seems to influence local acceptability of a proposed wind farm is "community benefits".

The definitions of both local community and community benefits are contested. The Centre for Sustainable Energy (CSE, 2009) observes that locality might be viewed differently when considering differing local benefits. For example, a wind turbine manufacturing factory can be viewed as having the benefit of offering jobs on a larger scale than just the community where it is situated. Furthermore, there is the question of whether local job gains are actually a community benefit. Local jobs could be a by-product of wind farm development, but they are often an unintended one and benefit only specific individuals and not the whole community. Therefore, the CSE (2009) only cite as community benefits those that can be purposefully disposed, and those that are supposed to benefit the whole local community rather than only some individuals. These include:

- community funds
- benefits in kind

- local ownership
- local contracting and local employment.

More recently, the Scottish Government took a different stance and listed the community benefits related to renewable energy developments as follows:

The local benefits arising from renewable energy developments can include:

1   Benefits derived from undertakings directly related to the development such as improved infrastructure,
2   Wider socio-economic community benefits in terms of job creation,
3   Benefits derived from community ownership in the development, referred to ... as "community investment",
4   Voluntary monetary payments to the community that are not related to anticipated impacts of the planning application usually provided via an annual cash sum, often referred to as a community benefit fund,
5   Other voluntary benefits which the developer provides to the community, (i.e. in-kind works, direct funding of projects, one-off funding, local energy discount scheme or any other site-specific benefits.).

(Scottish Government, 2013, p. 7)

"Community funds" or "community benefit funds" (whether following the CSE or Scottish Government definitions), are created by developers who pay a lump sum or make periodic payments to a fund that will benefit local residents' collective projects. In Scotland, the government has been promoting a standard practice which expects commercial developers to pay at least £5000 per year for each MW of installed capacity into a fund that benefits the local community, index-linked for the operational lifetime of the development (Scottish Government, 2013).

"Benefits in kind" are provided by developers who pay for local community facilities, environmental improvements or educational support, to name a few examples.

"Local ownership" or "community investment" (again differently labelled by the CSE and the Scottish Government) consists of ownership of shares by local citizens: project shares might be owned either through investment or through a profit-sharing or part-ownership scheme that ties the community benefit to the project's performance.

Finally, "local contracting and local employment" are purposefully set up by the developer during the construction and operational phases.

The "local ownership" category definition provided by CSE also seems to include co-operatives, along with any other scheme that allows local residents to invest in the project. While such a possibility is offered voluntarily by the developer, who could otherwise have pursued a standard commercial scheme, residents are nevertheless requested to invest in order to benefit from the scheme. Therefore, in this case, the project does not offer a benefit for the whole community because some residents might not invest; furthermore, an investment presupposes an element of risk, which in itself cannot be considered beneficial for residents/ investors. Thus, investment schemes should be distinguished from community benefits, even when the scheme is led by a developer who makes the "kind gesture" (or interested gesture) of opening it to local investment. After presenting the initial classification, CSE (2009) further specifies that, strictly speaking, such investment schemes do not fit their definition of community benefits because they are not delivered to the entire local community.

Different from the cases outlined above is a type of local ownership that is collective and which benefits the whole community. For example, a developer could benefit a local trust that represents the community by donating a number of shares in the local wind farm or the ownership of a turbine. The trust would benefit from the revenue generated by the wind farm in relation to the number of shares or turbine ownership, and the revenue could then be spent on projects that benefit the whole community. The Scottish Government (2013) includes "community ownership" within its list of local benefits.

In a separate document, the Scottish Government (2015, p. 15) lists three main options for "shared ownership" (ownership shared between the local community and the developer of a commercially-sized wind farm), pointing out that these options are flexible:

1   A joint venture vehicle can be set up, which will be part owned by both community group and developer. This may be referred to as a Special Purpose Vehicle (SPV). The community group will have the right to vote on the company's activities.

    • This is known as the "joint venture" model.

2   The developer owns the development (and may set up a new private company for this purpose), with the community buying the right to a defined percentage of revenues or net revenues (after operating costs and other costs have been paid). The community does not own any shares, so is not able to vote on the company's activities.

    • This is known as the "shared revenue" model.

3   The development is split into two and is owned discretely by the developer and the community group.

    • This is known as the "split ownership" model.

All three models allow for a co-operative scheme that would give residents the opportunity to invest and have a say according to the wind farm's co-operative governance model and the ownership model.

As has been made clear, no single definition of "community benefits" exists. Shared ownership can be seen as a community benefit or not, depending on how the ownership scheme is developed and whether it will benefit the whole community or just part of it. A co-operative model could have a role in certain types of shared ownership schemes that might divide the community between residents who invested and those who did not. Thus, it is perhaps more meaningful to think of the co-operative model as an opportunity for the community to join a shared ownership scheme and exert some control over it, rather than as a community benefit per se. If successful, a co-operative scheme could benefit the community; therefore, it is more of a "potential" rather than an actual community benefit.

Community benefits and social acceptability of renewable energy

Some studies have examined the role of community benefits in the social acceptability of renewable energy developments and wind energy in particular. Cass *et al.* (2010) conducted 49 semi-structured interviews, 34 focus groups and administered questionnaire surveys across 10 case studies of renewable energy developments in the UK. They found that developers' prime motivation for providing benefits was their desire to be seen as "good neighbours" and to re-localise the benefits deriving from the renewable energy development. However, developers were reluctant to adopt arguments that acknowledged the development's negative impacts and the right to compensation. Local people were divided, as a large number of individuals voiced suspicion or openly accused developers of attempting to bribe the local community. Moreover, those surveyed were often unable to give correct details of the community benefits specifically proposed for their local development. What emerged from the focus groups was a public interest in community benefits in the form of cheaper or free electricity, which some respondents wrongly assumed would be directly supplied by the local renewables development to the surrounding community buildings. Despite the scepticism towards community benefits, Cass *et al.* (2010) demonstrated a strong positive correlation between perceived benefits, at both a personal and a local level, and project support; furthermore, personal and local benefits influenced support more than any other variable they examined.

Similarly, Aitken (2010) researched the impact of community benefits on the acceptability of wind farms. She studied the case of a Scottish wind farm throughout all its phases—from planning to operation. The wind farm is comprised of 16 turbines for a total of 32 MW of installed capacity, and it is situated in between two small towns, one located 7 miles away and the other 3 miles away. The wind farm site stretches across two councils. Aitken carried out her research in three phases: a preliminary phase in which she collected and analysed written material of various sorts (including objection letters) regarding the then proposed wind farm; a second stage in which she observed the public inquiry that followed the initial rejection of the planning application and conducted a series of semi-structured interviews with the actors involved in the planning process; and a final phase that took place after the wind farm had been constructed. The latter involved individuals who had participated in the planning process and in the earlier round of interviews.

The developer announced the benefit scheme early on and it was sceptically received by those in the community who opposed the development, and they immediately pointed to the vagueness of the benefits provision or even viewed it as an attempt to bribe the community. Nevertheless, the letters of objection did not mention this issue very often. Interviews highlighted that, along with the critical appraisal of some objectors, others instead considered the offer of the developer as sensible. In order to avoid accusations of attempting to bribe the local community, the developer delayed the determination of the benefits package until after the final planning decision had been made. The benefits were eventually set out to be delivered in three forms: first, a fixed annual payment per MW of installed capacity; second, a variable annual payment linked to the production of energy; and third, a one-off payment of £75,000 for energy efficiency improvements in local community buildings. Controversies arose regarding the question of who belonged to the local community affected by the wind farm. Moreover, there were further disagreements regarding the composition of the panel of local community representatives who would be in charge of deciding which projects would be financed, and the rules that were chosen concerning how to assign the funding.

Aitken (2010) concludes that community benefits seen as compensation are a questionable strategy for increasing wind energy's social acceptability and that issues of trust and procedural justice must be carefully considered. She advocates for increased involvement of the local community in the planning process; in regard to community benefits, she suggests that these should be formalised by institutionalised guidance or rules which would make them less controversial and less likely to be considered as an "attempt to bribe" by communities.

Jones and Eiser's (2009) survey examined a specific form of community benefit, the "community trust fund". Their hierarchical regression analysis of the effects of economic benefit variables on support/opposition indicated that "community trust fund" was statistically significant for respondents from areas adjacent to a wind farm in influencing support/opposition, with a standardised beta coefficient of 0.10, which compared with coefficients of 0.32 for "general attitude to wind", 0.30 for "general economic benefit", 0.13 for "opportunity to invest and 0.13 for "cheaper electricity". It seems, therefore, that the possibility of a community fund influenced the support of respondents adjacent to the proposed wind sites, but to a lesser extent than other economic benefits.

The scholarly literature makes it clear that community benefits have a positive, albeit modest, impact on acceptability. Community benefits are often contentious, and they risk being considered a developer's tool for buying local consensus.

The co-operative model of local ownership

Co-operatives can be defined as: "independent, democratically controlled enterprises. They are owned and governed by their members, with the aim of meeting common social, economic and environmental needs." (Department of Trade and Industry and Co-operatives UK, 2005, p. 9).

In the context of wind energy, a co-operative model of wind farm community ownership allows local residents to buy shares in a local wind farm development and therefore benefit from the revenue produced by electricity generation and its sale.

The main focus of this work is to examine the co-operative model, which can be applied within different schemes of shared or full ownership of wind farms, in relation to the social acceptability of proposed wind farms. The reason for this choice lies in the specific characteristics of this model, and specifically in its democratic governance, which entails that each member has the right to one vote, irrespective of the number of shares that he or she might hold. Furthermore, as stated in definition above, a co-operative also has "common social and environmental" objectives beyond the merely economic goals pursued by regular companies; thus, it is particularly well placed to maximise the positive environmental and social outcomes of renewable energy generation.

Energy co-operatives are more common outside of the UK, although the country that has arguably employed the model the most, Denmark, has been using it less often since the 1990s (Sperling *et al.*, 2010). In fact, in Denmark, despite the country achieving a peak of 50 percent of total MW of installed wind capacity owned through co-operative schemes by 1994, this proportion fell considerably (to less than 20 per cent) by 2001 (Department of Trade and Industry, 2004). Sperling *et al.* (2010) attribute this decrease to a combination of causes: the increased size of wind turbines with consequent increased investment costs, the diffusion of offshore wind developments and higher income insecurity.

Co-operative ownership schemes have been introduced more recently in the UK, and have been applied to the development of commercial-sized wind farms, i.e. comprised of large, not isolated, wind turbines.[4] The UK co-operative model has been pioneered by Energy4All[5] and is often proposed as partial ownership of a commercial development. The model is realised through the creation of a local co-operative for the purpose of raising funds from local individual investors who are invited to buy shares in it (Scottish Government, 2009).

Energy4All was founded in 2002 but its story, according to the organisation, began in the mid-1990s "when an innovative Swedish company came to the UK to establish the sort of community ownership of wind farms that was already common in Sweden" (Energy4All, 2009, p. 2). The first community-owned co-operative in the UK was Baywind in Cumbria, a successful project whose revenue financed the creation of Energy4All. At the present time, the purpose of Energy4All is to promote the co-operative model of community ownership of wind farms in the UK by offering support to communities that express an interest in the scheme. Energy4All assists communities during both the pre-planning and post-planning phases. In the pre-planning phase, Energy4All helps the community identify sources of funding, a process termed "risk funding"; it also provides project management assistance and will work on disseminating information and engaging the local community. In the post-planning phase, depending on whether or not the project is entirely community owned, Energy4All can guide the procurement of turbines and other equipment, infrastructure contracts, grid connection, contracts for the sale of energy, etc. It can also oversee the whole construction of the wind farm (Energy4All, 2009). In the post-planning phase, a major step in the process is the creation of the co-operative and the issuing of a share offer which might need to be supplemented by obtaining a bank loan, if necessary, to complete the financing of the project.

Ownership of a wind farm in shared ownership cases is only partial, with the co-operative purchasing a stake in the future revenue of the project (Royalty Instrument Agreement), or through a joint venture with the commercial developer, or by eventually purchasing the ownership of one or more turbines in the development. Energy4All also supports other schemes: "100% community ownership"; the "regional co-operative model", in which a regional or national co-operative covering a wide geographic area and investing in several projects finances the project; and, the "loan model", in which a community can approach an existing energy co-operative to arrange a loan to get another project started.

Shares are set at an affordable price (e.g. for the Energy4All Westmill Wind Farm co-operative scheme the minimum investment was £250 and the maximum investment, per individual, was £20,000). Interest is paid annually to shareholders. While members might hold different numbers of shares, the co-operative operates on the principle of "one member, one vote". Members have the right to elect the board that runs the co-operative, and every year a third of the board has to step down, although they are eligible for re-election.

Toke (2002) assumes the effects of the co-operative model on local wind farm acceptability to be positive (i.e. the scheme would increase local acceptability). He believes that this model has made wind energy more popular in the recent past in Denmark and notes a report by Olesen (1998): "One of the most important lessons from the Danish wind turbine market is that wherever co-operatives with local ownership were involved, the local population is much more in favour of the project" (cited in Toke, 2002, p. 93). In another article, Toke (2005a, p. 302) states:

> Selling shares to the general public (with preference given to local people) allows positive public relations. [...] It also has broader strategic political advantages for wind power. It creates a group of people who, having made a significant personal commitment, are likely to campaign in support of wind power.

More recently, others (Lipp and McMurtry, 2015; Bauwens *et al.*, 2016; Bauwens and Devine-Wright, 2018) have reiterated Toke's arguments. Bauwens (2015) published a study in which he surveyed 222 members of a West Flanders wind farm co-operative called BeauVent and an equal number of non-members who resided in Flanders. Using a matching technique, the propensity to accept wind energy in general and wind farms in the locale was estimated, resulting in a statistically significant increase of about 7 per cent for co-operative members compared with the non-member group. In a larger study conducted in Flanders, Bauwens and Devine-Wright (2018) surveyed nearly 4000 individuals and contrasted the answers provided by members of two different types of renewable energy co-operative—BeauVent, a wind farm, and Ecopower, a large producer of renewable energy from various sources as well as an electricity utility provider—with those of non-members. The authors found that co-operative members were more supportive of renewable energy, onshore developments and local wind farms than non-members, and that the latter had a higher frequency of "neither agree nor disagree" answers. This showed that non-members were more indifferent than co-operative members, rather than being against renewables developments; this, changed, however, when respondents were asked specifically about wind energy, which elicited more polarised views between co-operative members and non-members.

While wind farm co-operatives might increase the acceptability of wind farms, the evidence is very limited. The only two empirical studies traced (Bauwens, 2015; Bauwens and Devine-Wright, 2018) targeted, on the one hand, members of co-operatives already in place (whose perceptions might have changed over time compared with the more delicate planning stage) and, on the other, members of the general public who might not have faced the dilemma of whether to support or oppose a wind farm in their locale.

Non-co-operative local ownership investment schemes

Several other local ownership schemes that involve the investment of some individuals, or of the entire community, can positively influence wind farm acceptability.

A well-known example of non-co-operative local ownership of a commercial-sized development is the case of Fintry in Stirlingshire. In 2003, in the village of Fintry, villagers created the Fintry Renewable Energy Enterprise (FREE) with the aim of making Fintry a carbon-neutral, sustainable community. They viewed the wind farm to be developed at Earlsburn by Falk Renewables as an opportunity, and they asked the developer to agree to a scheme which would allow the village to own one of the development's wind turbines. The developer agreed, financing the additional turbine through a bank loan that would have financed the whole wind farm. The agreement with FREE was that the village would repay the portion of the loan dedicated to the additional wind turbine over the wind farm's 15 years of operation (CSE, 2009; Scottish Government, 2009).

Another three cases of non-co-operative local ownership were researched by Maruyama *et al.* (2007), who investigated the motives behind the choice of whether or not to invest in community-funded wind farm developments in Japan. The wind farms under consideration were small and consisted of single turbine developments, with outputs between 1 and 1.5 MW, and sited in three locations (Hokkaido, Aomori and Akita). The authors surveyed four groups of individuals: investors in the Hokkaido wind turbine; investors in the Aomori prefecture wind turbine; investors in the Japan Green Fund (whose finances would support the turbines in the Akita and Aomori prefectures); and finally, a group of non-investors who had initially expressed an interest in the Aomori turbine or in the national fund. Written questionnaires were used to survey respondents in 2003 and 2005. Factor analysis was performed on 11 questions that aimed to clarify respondents' interest in the wind community developments. Three factors were identified as influential. The first factor, "environmental movement", aggregates the beliefs that collective environmental action is needed; the second, "commitment factor", expresses the wish to participate in community-owned wind energy, but limited only to economic participation, which is easier than actively engaging in an environmental movement. Finally, the third, "economic factor", expresses the desire for economic participation that is not a donation and that delivers a dividend.

A multivariate analysis of variance (MANOVA) was performed on the average scores of the three factors used as dependent variables. The results showed that the second factor, which expressed the desire to participate in community-owned wind projects, was relatively high among investors in the Aomori turbine, but not among investors in the Hokkaido turbine or in the Japan Green Fund. In both these two last groups of investors, the factor having the highest value was "environmental movement", which had a negative value in the Aomori case. Maruyama *et al.* (2007) explain the differences in terms of variations in the social contexts of the three collective funding initiatives, which present different levels of pro-environmental culture and social cohesion that are suggested to explain the differences across the cases.

Another study by Warren and McFadyen (2010) compares the Scottish cases of the Isle of Gigha and the peninsula of Kintyre in the Argyll and Bute council area. The Isle of Gigha was bought by its community in 2002 and is now owned and managed by a development trust, the Gigha Heritage Trust. Islanders decided in 2005 to invest in wind energy and set up a small wind farm comprising three pre-commissioned 225 kW wind turbines which amount to a total installed capacity output of nearly 0.7 MW. The scheme was financed through a mix of grant funding, loan finance and equity finance (The Isle of Gigha Heritage Trust, n.d.). The Isle of Gigha case was compared with the nearby case of Kintyre where, at the time of the study, three commercial wind farms were operational and two were under construction. The size of the Gigha and Kintyre developments are radically different: the total capacity installed at the time of the survey on the Kintyre peninsula was 58.6 MW, which would have risen to 102.6 MW once the two wind farms under construction had been completed. Therefore, the visual impact on Kintyre was significantly higher than on Gigha.

Warren and McFadyen (2010) surveyed Gigha and Kintyre residents and tourists using a questionnaire with 106 respondents and supplemented these with face-to-face interviews with 5 stakeholders. Local residents accounted for 68 of the responses (24 were based in Gigha and 44 in Kintyre), while tourists accounted for 38 of the responses. The questionnaires were partly administered face-to-face (61 per cent) and partly completed online (39 per cent). The results indicate that the Gigha respondents were consistently more positive about wind power in general and about the local wind developments. Among Gigha residents, 96 per cent supported increasing wind energy developments in Scotland versus 68 per cent in Kintyre. Residents in Gigha also showed more support for increasing the number of wind farms in their local area (75 per cent versus 64 per cent for Kintyre); more opposed this option in Kintyre (12 per cent) than in Gigha (8 per cent). The authors did not provide correlation statistics to ascertain whether the area of belonging (Gigha or Kintyre) influenced support for wind in general or for additional wind farms in the local area. Moreover, the authors themselves admit that they compared sharply contrasting local areas in terms of size of wind farm developments. Their choice of using non-probabilistic sampling was also problematic: respondents to the online questionnaire would have been self-selected.

Jones and Eiser (2009) researched four proposed wind farms in the area of Sheffield and considered "opportunity to invest" within a set of variables that could possibly lead to local support for/opposition to wind farms. Their regression analysis of the variables pertaining to positive economic benefits, controlling for the "general attitude to wind" variable, shows that the variable "opportunity to invest" had a positive effect in influencing support, with a standardised beta coefficient of 0.13. This is nevertheless smaller in magnitude than the effect of another statistically significant economic benefit variable, "general economic benefit" (0.30), while it compares equally with "cheaper electricity" (0.13) and shows a greater effect than "community trust fund" (0.10).

From the limited research reviewed, it appears that community-owned energy projects have a positive impact on the community and could possibly increase the acceptability of wind energy. Again, it has to be assumed that local investment by residents to raise funds, if necessary, might be a divisive process; however, other opportunities to raise funds can be pursued, particularly in the case of small developments.

The case for co-operatives, producers of green electricity

Owen and Hunt (2004) consider community involvement to be essential to achieving the carbon dioxide reduction goal set by the UK government. Focusing on the UK context, the authors highlight four reasons for local ownership of renewable energy developments:

1 "Local ownership creates local dialogue and acceptance."
2 "Local ownership raises public awareness and makes sustainable development understandable."
3 "Local ownership solves problems and conflicts."
4 "Local ownership gives people the opportunity to act for sustainable development." (Owen and Hunt, 2004, p. 5)

The authors also consider the barriers to achieving this goal in communities throughout the country, identifying "complexity" as the major impediment.

Complexity is implicated in several problems: raising funds is difficult because financial support, if available, is not easily identifiable because of the large number of schemes; securing planning permission is hard to achieve; and the amount of upfront funding needed is a major issue. All of these difficulties make the process costly and time-consuming.

Bauwens *et al.* (2016) also draw attention to barriers to wind farm co-operatives in their analysis of the institutional environment of wind farm co-operatives in four European countries (Belgium, Denmark, Germany and the UK). The authors point to the complexities of the planning system, the limitedness of incentives and the contextual tendency to adopt a larger industry standard for wind turbines, all of which have provoked more stringent regulatory requirements.

Toke (2002) also considers the importance of community involvement in renewable energy development. He uses a rational choice model of collective action first outlined by Olson (1965) to explain why co-operative wind power schemes were well developed in Denmark, while wind met difficulties in expanding in the UK. According to Toke, Danish policymakers privileged co-operative schemes, while in the UK it was more convenient for large commercial companies to invest in wind power. This difference led to strong local opposition to wind farms in the UK, because they were private interest schemes that did not involve local people, while collective, co-operative schemes were widely accepted in Denmark. When, in the last few years, individual local ownership of wind farms expanded in Denmark, there was increased criticism towards wind farm developments that was attributed to the lack of collective local ownership (Toke, 2002).

Toke's rational choice model describes wind farm protesters as an example of "free riders", individuals who have a positive attitude towards the environment and renewable energy but do not want to bear the costs (such as noise and visual impact) of wind farms, even if, at the same time, they are willing to "free ride" the benefits, (e.g. a less polluted environment) when the wind farms are situated in other locations. The author holds that co-operative schemes have been shown to be effective in realising broader local acceptance of wind developments, balancing, for the individuals involved, the costs of wind farms (noise and visual impact) with the advantages of local ownership (economic and other selective incentives). Hence, Toke's rational choice model suggests a solution, at least in the specific case of wind farms opposition, for the "value action gap" described in research (Kollmuss and Agyeman, 2002)—the gap between environmental behaviours and actors' pro-environmental attitudes.

A proposed framework of participation in green electricity co-operatives

In order to better describe which variables might influence participation in community wind farm co-operatives, the literature review presented in this chapter was used to assist in drafting a theoretical framework which includes several variables that affect participation in such co-operatives. The categorisation in Figure 2.1 is not designed to be exhaustive, but rather to serve as a basis for the first stage of data collection that was carried out in this research project.

Drawing on Stern (2000), three categories of variables, which have been discussed in this chapter, are considered capable of influencing local acceptance of wind farms and participation in wind farm co-operatives: "contextual variables", "personal capabilities" and "psychological variables" (see Figure 2.1).

It has been proposed (Pellegrini-Masini, 2007) that the effectiveness of pro-environmental attitudes in shaping behaviour only in low-cost situations (Diekmann and Preisendörfer, 2003) is indicative of the presence of a preeminent hierarchy of motivations that are likely to reflect a hierarchy of needs of human agency. This hierarchy distinguishes primary (physiological, safety and belongingness) and secondary needs (esteem and self-actualization), as originally proposed by Maslow (1987) and empirically supported by several studies (Oishi *et al.*, 1999; Sheldon *et al.*, 2001; Taormina and Gao, 2013). A hierarchy of needs could hence explain why pro-environmental attitudes might be lacking effectiveness in driving pro-environmental behaviours in the presence of subjectively perceived high costs. Ultimately, this hierarchy could orientate the priorities of individuals who would, therefore, perceive differently costs and

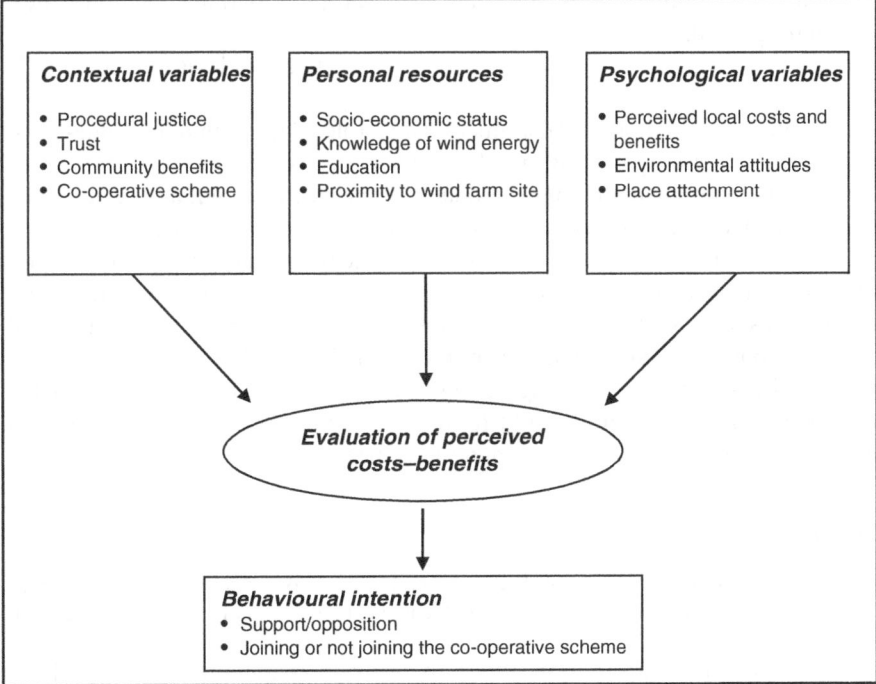

Figure 2.1 Theoretical framework of wind farm acceptance and participation in community co-operative scheme.

benefits concerning a hypothetical behaviour, such as supporting or opposing a proposed wind farm or joining a wind farm co-operative.

The influence of costs and benefits on participatory behaviours and environmentally responsible behaviours (ERBs) is widely acknowledged (e.g. Whiteley and Seyd, 1996; Stern, 2000; Diekmann and Preisendörfer, 2003; Simmons and Birchall, 2003). Therefore, following Stern (2000), "perceived costs" and "perceived benefits" are considered as psychological variables in the theoretical framework outlined in this section. The adoption of the labels "perceived costs" and "perceived benefits" instead of "costs" and "incentives" underlines the assumption that what ultimately influences behavioural intentions are neither objective "costs" nor objective "benefits", but rather the subjective perception of these costs and benefits. To explain individuals' choices of whether to engage in ERBs, the significance of their perceived benefits and costs was indicated:

> We argue that acting a pro-environmental behaviour is the result of an individual choice which follows a subjective cost–benefit analysis. This cost–benefit analysis is influenced primarily by the goal orientation of an individual at a certain point in time, which is in turn ultimately determined by the level of satisfaction of their needs. Secondarily by the availability of financial resources and the level of knowledge and information that an individual holds about the best course of action to achieve their main goal(s).
>
> (Pellegrini-Masini, 2007, p. 3)

Therefore pro-environmental attitudes play a role, along with other value-based assessments regarding the cost and benefits of a considered behaviour, because acting pro-environmentally might produce the benefit of avoiding cognitive dissonance with one's pro-environmental attitudes (Thøgersen, 2004).

The pre-eminence of perceived costs and benefits in determining the acceptability of renewable energy installations has been more recently supported by a comprehensive literature review. This stresses the relevance of contextual factors, along with environmental attitudes, in determining the acceptance of energy alternatives, thereby suggesting that this approach might be a viable theoretical standpoint which could further the explanation of the acceptance of energy infrastructure (Perlaviciute and Steg, 2014).

Notes

1 The authors limit themselves to defining it as follows: "A concern that wind development would introduce general unwanted change to the community ..." (Jones and Eiser, 2009, p. 4611).
2 "To deliver on its targets WAG [Welsh Assembly Government] planning guidance identified seven Strategic Search Areas ... suitable for large-scale wind energy development, amounting to a significant concentration of development potential on specific, remote areas of rural Wales" (Munday *et al.*, 2011, p. 2).

3 The acronym NIMBY, i.e. "not in my back yard", is defined as "a colloquialism signifying one's opposition to the locating of something considered undesirable in one's neighborhood" (Kinder, 2016).
4 While there is no classification of wind farms based on size, the Scottish Government (2009) considers commercial farms to be of 20 MW or more.
5 https://energy4all.co.uk/.

References

Agterbosch, S., Meertens, R. M. and Vermeulen, W. J. V. 2009. The relative importance of social and institutional conditions in the planning of wind power projects. *Renewable and Sustainable Energy Reviews*, 13, 393–405.

Aitken, M. 2010. Wind power and community benefits: Challenges and opportunities. *Energy Policy*, 38, 6066–6075.

Awerbuch, S. 2006. Portfolio-based electricity generation planning: Policy implications for renewables and energy security. *Mitigation and Adaptation Strategies for Global Change*, 11, 693–710.

Bamberg, S. and Möser, G. 2007. Twenty years after Hines, Hungerford, and Tomera: A new meta-analysis of psycho-social determinants of pro-environmental behaviour. *Journal of Environmental Psychology*, 27, 14–25.

Barr, S., Gilg, A. W. and Ford, N. 2005. The household energy gap: Examining the divide between habitual- and purchase-related conservation behaviours. *Energy Policy*, 33, 1425–1444.

Barry, J. and Proops, J. 1999. Seeking sustainability discourses with Q methodology. *Ecological Economics*, 28, 337–345.

Bauwens, T. 2015. Propriété coopérative et acceptabilité sociale de l'éolien terrestre. *Reflets et perspectives de la vie économique*, LIV, 59–70.

Bauwens, T. and Devine-Wright, P. 2018. Positive energies? An empirical study of community energy participation and attitudes to renewable energy. *Energy Policy*, 118, 612–625.

Bauwens, T., Gotchev, B. and Holstenkamp, L. 2016. What drives the development of community energy in Europe? The case of wind power cooperatives. *Energy Research and Social Science*, 13, 136–147.

Bellhouse, G. 2004. *Low frequency noise and infrasound from wind turbine generators: A literature review.* Available at: http://canwea.ca/pdf/talkwind/Low%20Frequency%20Noise%20and%20Infrasound%20from%20Wind%20Turbine%20Generators.pdf. [Accessed 19 December 2016].

Bishop, I. D. and Miller, D. R. 2007. Visual assessment of off-shore wind turbines: The influence of distance, contrast, movement and social variables. *Renewable Energy*, 32, 814–831.

Bonaiuto, M., Carrus, G., Martorella, H., Bonnes, M. and Mirilia, B. 2002. Local identity processes and environmental attitudes in land use changes: The case of natural protected areas. *Journal of Economic Psychology*, 23, 631–653.

Braunholtz, S. and McWhannel, F. 2003. *Public attitudes to windfarms: A survey of local residents in Scotland.* MORI Scotland, Edinburgh. Available at: www.scotland.gov.uk/socialresearch.

Cass, N., Walker, G. and Devine-Wright, P. 2010. Good neighbours, public relations and bribes: The politics and perceptions of community benefit provision in renewable energy development in the UK. *Journal of Environmental Policy & Planning*, 12, 255–275.

Country Guardian. 2000. *The case against wind farms* [Online]. Available at: www. digknow.com/pdf/The%20Case%20Against%20Windfarms.pdf [Accessed 28 March 2011].

CSE (Centre for Sustainable Energy). 2005. *Community benefits from wind power: A study of UK practice & comparison with leading European countries*. Department of Trade and Industry, London. Available at: www.cse.org.uk/pdf/pub1049.pdf.

CSE (Centre for Sustainable Energy). 2009. *Delivering community benefits from wind energy development: A toolkit*. Department of Trade and Industry, London. Available at: www.cse.org.uk/downloads/toolkits/community-energy/planning/renewables/delivering-community-benefits-from-wind-energy-tookit.pdf.

Dalton, G. J., Lockington, D. A. and Baldock, T. E. 2008. A survey of tourist attitudes to renewable energy supply in Australian hotel accommodation. *Renewable Energy*, 33, 2174–2185.

Department of Trade and Industry. 2004. *Co-operative energy: Lessons from Denmark and Sweden*. Department of Trade and Industry, London.

Department of Trade and Industry and Co-operatives UK. 2005. *Co-operative energy: Lessons from Denmark and Sweden*. Department of Trade and Industry, London. Available at: www.oti.globalwatchonline.com/online_pdfs/36247MR.pdf?pubpdfd load=05%2F592.

Devine-Wright, P. 2005a. Beyond NIMBYism: Towards an integrated framework for understanding public perceptions of wind energy. *Wind Energy*, 8, 125–139.

Devine-Wright, P. 2005b. Local aspects of UK renewable energy development: Exploring public beliefs and policy implications. *Local Environment*, 10, 57–69.

Devine-Wright, P. 2008. Reconsidering public acceptance of renewable energy technologies: A critical review. In: Jamasb, T., Grubb, M. and Pollitt, M. (eds.) *Delivering a low carbon electricity system: Technologies, economics and policy*. Cambridge: Cambridge University Press.

Devine-Wright, P. 2009. Rethinking NIMBYism: The role of place attachment and place identity in explaining place-protective action. *Journal of Community & Applied Social Psychology*, 19, 426–441.

Devine-Wright, P. 2013. Explaining "NIMBY" objections to a power line: The role of personal, place attachment and project-related factors. *Environment and Behavior*, 45, 761–781.

Devine-Wright, P. and Devine-Wright, H. 2006. Social representations of intermittency and the shaping of public support for wind energy in the UK. *International Journal of Global Energy Issues*, 25, 243–256.

Devine-Wright, P. and Howes, Y. 2010. Disruption to place attachment and the protection of restorative environments: A wind energy case study. *Journal of Environmental Psychology*, 30, 271–280.

Devine-Wright, P. and Wiersma, B. 2020. Understanding community acceptance of a potential offshore wind energy project in different locations: An island-based analysis of "place-technology fit". *Energy Policy*, 137, 111086. Available at https://doi.org/10.1016/j.enpol.2019.111086.

Diekmann, A. and Preisendörfer, P. 2003. Green and greenback. *Rationality and Society*, 15, 441–472.

Dimitropoulos, A. and Kontoleon, A. 2009. Assessing the determinants of local acceptability of wind-farm investment: A choice experiment in the Greek Aegean Islands. *Energy Policy*, 37, 1842–1854.

Dudleston, A. 2000. *Public attitudes towards wind farms in Scotland: Results of a residents survey.* The Scottish Executive, Edinburgh.

Ellis, G., Barry, J. and Robinson, C. 2007. Many ways to say no, different ways to say yes: Applying Q-Methodology to understand public acceptance of wind farm proposals. *Journal of Environmental Planning and Management,* 50, 517–551.

Eltham, D. C., Harrison, G. P. and Allen, S. J. 2008. Change in public attitudes towards a Cornish wind farm: Implications for planning. *Energy Policy,* 36, 23–33.

Energy4All. 2009. *Clean, green energy—Let's cooperate* [Online].

Exeter Enterprises. 1994. *Public attitudes towards wind power—A survey of opinion in Cornwall and Devon.* Department of Trade and Industry, London.

Firestone, J. and Kempton, W. 2007. Public opinion about large offshore wind power: Underlying factors. *Energy Policy,* 35, 1584–1598.

Firestone, J., Kempton, W. and Krueger, A. 2009. Public acceptance of offshore wind power projects in the USA. *Wind Energy,* 12, 183–202.

Frantál, B. and Kunc, J. 2011. Wind turbines in tourism landscapes: Czech experience. *Annals of Tourism Research,* 38, 499–519.

Glasgow Caledonian University, Moffat Centre and Cogentsi. 2008. *The economic impact of wind farms on Scottish Tourism: A report for the Scottish Government.*: Energy and Climate Change Directorate, Scottish Government, Edinburgh. Available at: www.scotland.gov.uk/Resource/Doc/214910/0057316.pdf.

Gross, C. 2007. Community perspectives of wind energy in Australia: The application of a justice and community fairness framework to increase social acceptance. *Energy Policy,* 35, 2727–2736.

Haggett, C. 2004. *Tilting at windmills? The attitude–behaviour gap in renewable energy conflicts.* Landscape Research Group, University of Newcastle, UK. Available at: www.psi.org.uk/ehb/docs/finalreport-Haggett.pdf.

Hidalgo, M. C. and Hernández, B. 2001. Place attachment: Conceptual and empirical questions. *Journal of Environmental Psychology,* 21, 273–281.

Hoepman, N. 1998. Four de Wyn, Provinsje Friesland.

Janssen, S. A., Eisses, A. R., Vos, H. and Pedersen, E. 2009. Exposure-response relationships for annoyance by wind turbine noise: A comparison with other stationary sources. Eighth European Conference on Noise Control 2009 (EURONOISE 2009)— Proceedings of the Institute of Acoustics, 26–28 October, Institute of Acoustics (Great Britain), Edinburgh.

Jones, C. R. and Eiser, J. R. 2009. Identifying predictors of attitudes towards local onshore wind development with reference to an English case study. *Energy Policy,* 37, 4604–4614.

Kinder, P. D. 2016. *Not in my backyard phenomenon (NIMBY)* [Online]. *Encyclopædia Britannica.* Available at: www.britannica.com/topic/Not-in-My-Backyard-Phenomenon [Accessed 20 April 2011].

Klick, H. and Smith, E. R. A. N. 2010. Public understanding of and support for wind power in the United States. *Renewable Energy,* 35, 1585–1591.

Kollmuss, A. and Agyeman, J. 2002. Mind the gap: Why do people act environmentally and what are the barriers to pro-environmental behavior? *Environmental Education Research,* 8, 239–260.

Krohn, S. and Damborg, S. 1999. On public attitudes towards wind power. *Renewable Energy,* 16, 954–960.

Lipp, J. and McMurtry, J. 2015. *Benefits of renewable energy co-operatives: Summary of literature review from the Measuring the Co-operative Difference Research Network.*

Measuring the Co-operative Difference Research Network. Available at: www.cooperativedifference.coop/wp-content/uploads/2015/02/Benefits-of-Renewable-Energy-Co-ops.pdf.

Lothian, A. 2008. Scenic perceptions of the visual effects of wind farms on South Australian landscapes. *Geographical Research*, 46, 196–207.

Maruyama, Y., Nishikido, M. and Iida, T. 2007. The rise of community wind power in Japan: Enhanced acceptance through social innovation. *Energy Policy*, 35, 2761–2769.

Maslow, A. H. 1987. *Motivation and personality*. New York: Harper & Row.

McLaren Loring, J. 2007. Wind energy planning in England, Wales and Denmark: Factors influencing project success. *Energy Policy*, 35, 2648–2660.

Meyerhoff, J., Ohl, C. and Hartje, V. 2010. Landscape externalities from onshore wind power. *Energy Policy*, 38, 82–92.

MORI Scotland. 2002. *Tourist attitudes towards wind farms*. Research study conducted for Scottish Renewables Forum and the British Wind Energy Association, summary report, MORI Scotland, Edinburgh.

Munday, M., Bristow, G. and Cowell, R. 2011. Wind farms in rural areas: How far do community benefits from wind farms represent a local economic development opportunity? *Journal of Rural Studies*, 27(1), 1–12.

Oishi, S., Diener, E. F., Lucas, R. E., Suh, E. M., Diener, E. F., Lucas, R. E. and Suh, E. M. 1999. Cross-cultural variations in predictors of life satisfaction: Perspectives from needs and values. *Personality and Social Psychology Bulletin*, 25, 980–990.

Olesen, G. 1998. *Large scale implementation of renewable and sustainable energy*. Danish Organisation for Renewable Energy, Arhus, Denmark.

Olson, M. 1965. *Logic of collective action public goods and the theory of groups* (revised edn.). London: Harvard University Press.

Owen, G. and Hunt, P. 2004. *Community engagement in energy through energy mutuals*. London: Mutuo.

Pedersen, E. and Larsman, P. 2008. The impact of visual factors on noise annoyance among people living in the vicinity of wind turbines. *Journal of Environmental Psychology*, 28, 379–389.

Pedersen, E. and Persson-Waye, K. 2004. Perception and annoyance due to wind turbine noise: A dose–response relationship. *Journal of the Acoustical Society of America*, 116, 3460–3470.

Pedersen, E. and Persson-Waye, K. 2007. Wind turbine noise, annoyance and self-reported health and well-being in different living environments. *Occupational and Environmental Medicine*, 64, 480–486.

Pedersen, E. and Persson-Waye, K. 2008. Wind turbines—low level noise sources interfering with restoration? *Environmental Research Letters*, 3, 015002.

Pedersen, E., Hallberg, L.-M. and Waye, K. P. 2007. Living in the vicinity of wind turbines: A grounded theory study. *Qualitative Research in Psychology*, 4, 49–63.

Pellegrini-Masini, G. 2007. The carbon-saving behaviour of residential households. Futures of Cities–51st IFHP World Congress, 23–26 September 2007, Copenhagen.

Perlaviciute, G. and Steg, L. 2014. Contextual and psychological factors shaping evaluations and acceptability of energy alternatives: Integrated review and research agenda. *Renewable and Sustainable Energy Reviews*, 35, 361–381.

Roberts, M. and Roberts, J. 2009. *Evaluation of the scientific literature on the health effects associated with wind turbines and low frequency sound*. Exponent, Wood Dale, IL.

Scottish Government. 2009. *Community renewable energy toolkit*. Scottish Government, Edinburgh.

Scottish Government. 2013. *Good practice principles for community benefits from onshore renewable energy developments.* Energy and Climate Change Directorate, Scottish Government, Edinburgh. Available at: www2.gov.scot/resource/0043/0043 8782.pdf.

Scottish Government. 2015. *Good practice principles for shared ownership of onshore renewable energy developments.* Energy Saving Trust, Edinburgh. Available at: www. localenergyscotland.org/media/79714/Shared-Ownership-Good-Practice-Principles.pdf.

SEI. 2003. *Attitudes towards the development of wind farms in Ireland.* Renewable Energy Information Office, Bandon. Ireland. Available at: www.seai.ie/uploadedfiles/ RenewableEnergy/Attitudestowardswind.pdf.

Sheldon, K. M., Elliot, A. J., Kim, Y. and Kasser, T. 2001. What is satisfying about satisfying events? Testing 10 candidate psychological needs. *Journal of Personality and Social Psychology,* 80(2), 325–339.

Shen, J. and Saijo, T. 2008. Reexamining the relations between socio-demographic characteristics and individual environmental concern: Evidence from Shanghai data. *Journal of Environmental Psychology,* 28, 42–50.

Simmons, R. and Birchall, J. 2003. Bringing citizens back into public services: Strengthening the "Participation Chain". ECPR Joint Sessions, Edinburgh, 28 March–2 April 2003.

Sims, S. and Dent, P. 2007. Property stigma: Wind farms are just the latest fashion. *Journal of Property Investment & Finance,* 25, 626–651.

Sims, S., Dent, P. and Oskrochi, G. 2008. Modelling the impact of wind farms on house prices in the UK. *International Journal of Strategic Property Management,* 12, 251–269.

Spence, A., Venables, D., Pidgeon, N., Poortinga, W. and Demski, C. 2010. *Public perceptions of climate change and energy futures in Britain: Summary findings of a survey conducted from January to March 2010.* Understanding Risk Working Paper 10-01, School of Psychology, Cardiff University. Available at: https://sp.ukdataservice.ac.uk/ doc/6581/mrdoc/pdf/6581final_report.pdf.

Sperling, K., Hvelplund, F. and Mathiesen, B. V. 2010. Evaluation of wind power planning in Denmark: Towards an integrated perspective. *Energy,* 35, 5443–5454.

Stern, P. C. 2000. New environmental theories: Toward a coherent theory of environmentally significant behavior. *Journal of Social Issues,* 56, 407–424.

Stern, P. C., Dietz, T., Abel, T. D., Guagnano, G. A. and Kalof, L. 1999. A value-belief-norm theory of support for social movements: The case of environmentalism. *Human Ecology Review,* 6, 81–97.

Sterzinger, G., Beck, F. and Kostiuk, D. 2003. *The effect of wind development on local property values.* Washington: Renewable Energy Policy Project.

Sustainable Development Commission. 2005. *Wind power in the UK: A guide to the key issues surrounding onshore wind power development in the UK.* Sustainable Development Commission, London. Available at: www.sd-commission.org.uk/data/files/ publications/Wind_Energy-NovRev2005.pdf.

Swofford, J. and Slattery, M. 2010. Public attitudes of wind energy in Texas: Local communities in close proximity to wind farms and their effect on decision-making. *Energy Policy,* 38, 2508–2519.

Taormina, R. J. and Gao, J. H. 2013. Maslow and the motivation hierarchy: Measuring satisfaction of the needs. *American Journal of Psychology,* 126, 155–177.

The Isle of Gigha Heritage Trust. n.d. *Frequently asked questions about the Gigha Windmills* [Online]. Available at: www.gigha.org.uk/windmills/TheStoryoftheWindmills.php [Accessed 2 March 2011].

Thøgersen, J. 2004. A cognitive dissonance interpretation of consistencies and inconsistencies in environmentally responsible behavior. *Journal of Environmental Psychology*, 24, 93–103.

Toke, D. 2002. Wind power in UK and Denmark: Can rational choice help explain different outcomes? *Environmental Politics*, 11, 83–100.

Toke, D. 2005a. Community wind power in Europe and in the UK. *Wind Engineering*, 29, 301–308.

Toke, D. 2005b. Explaining wind power planning outcomes: Some findings from a study in England and Wales. *Energy Policy*, 33, 1527–1539.

Toke, D., Breukers, S. and Wolsink, M. 2008. Wind power deployment outcomes: How can we account for the differences? *Renewable and Sustainable Energy Reviews*, 12, 1129–1147.

van Der Horst, D. 2007. NIMBY or not? Exploring the relevance of location and the politics of voiced opinions in renewable energy siting controversies. *Energy Policy*, 35, 2705–2714.

van Der Horst, D. and Toke, D. 2010. Exploring the landscape of wind farm developments: Local area characteristics and planning process outcomes in rural England. *Land Use Policy*, 27, 214–221.

Vorkinn, M. and Riese, H. 2001. Environmental concern in a local context. *Environment and Behavior*, 33, 249–263.

Warren, C. R. and McFadyen, M. 2010. Does community ownership affect public attitudes to wind energy? A case study from south-west Scotland. *Land Use Policy*, 27, 204–213.

Warren, C., Lumsden, C., O'Dowd, S. and Birnie, R. 2005. "Green on green": Public perceptions of wind power in Scotland and Ireland. *Journal of Environmental Planning and Management*, 48, 853–875.

Whiteley, P. and Seyd, P. 1996. Rationality and party activism: Encompassing tests of alternative models of political participation. *European Journal of Political Research*, 29, 215–234.

Williams, A. 1993. Role of fossil fuels in electricity generation and their environmental impact. *Science, Measurement and Technology*, 140, 8–12.

Wind Europe. 2019. *Wind energy in Europe in 2018. Trends and statistics*. Available at: https://windeurope.org/wp-content/uploads/files/about-wind/statistics/WindEurope-Annual-Statistics-2018.pdf [Accessed 11 March 2020].

Wolsink, M. 2000. Wind power and the NIMBY-myth: Institutional capacity and the limited significance of public support. *Renewable Energy*, 21, 49–64.

Wolsink, M. 2007a. Planning of renewables schemes: Deliberative and fair decision-making on landscape issues instead of reproachful accusations of non-cooperation. *Energy Policy*, 35, 2692–2704.

Wolsink, M. 2007b. Wind power implementation: The nature of public attitudes: Equity and fairness instead of "backyard motives". *Renewable and Sustainable Energy Reviews*, 11, 1188–1207.

Young, B. 1993. *Attitudes towards wind power: A Survey of opinion in Cornwall and Devon*. Energy Technology Support Unit, Harwell, UK.

3 Testing the theory
Methods and data collection

Research questions

The co-operative model seems to be a suitable instrument for achieving democratic engagement in the energy sector. This engagement could be viewed as part of the wider renewal of the democratic system through citizen engagement in associations advocated by Hirst (1994, 2002). Dobson (2003) explored the import of the environment in relation to the concept of citizenship. The construct of citizenship ultimately regulates the interaction between citizens and public goods (one such good being the environment) through the balance of citizens' rights and duties. In modern democracies, therefore, the environment emerges as an arena for democratic public debate.

In this context, Rifkin (2002) sees energy co-operatives as a means to facilitate public control over the energy system, therefore leading to a better, more equal distribution of wealth. In the specific context of wind power, Toke (2002) suggests that wind energy co-operatives could reduce and possibly even overcome opposition to hosting a wind farm in a local vicinity.

The present research aims to fill gaps in the literature regarding wind farms by addressing four central research questions:

1   Which factors influence the acceptability of wind farms? How do they relate to one another?
2   Which factors influence participation in wind farm co-operatives? How do they relate to one another?
3   Is the co-operative model effective in eliciting the participation of local communities and in overcoming opposition to wind developments? Why?
4   Do people perceive their status as citizens as a source of moral obligation to protect the environment? (In other words, does environmental citizenship denote a moral obligation to protect the environment?)

The scientific community has left research questions 2, 3 and 4 above largely unaddressed. While recent research sheds some light on the wider phenomenon of community energy projects (Walker, 2007) and on community ownership of wind farms (Warren and McFadyen, 2010), research on wind farm co-operatives is limited and primarily focuses on citizens who are already members of established wind farm co-operatives (Bauwens and Devine-Wright, 2018) rather than on proposed co-operatives. A handful of studies address how the concept of citizenship relates to the debate on wind farms and their acceptability. However, while a number of scholars have explored the factors that influence wind energy's acceptability, they have limited their focus to specific factors and have thus failed to provide a comprehensive explanation of opposition to local wind farms.

Research design

This research utilised a multi-method approach called triangulation (Jick, 1979). Denzin (1978) defined triangulation as "the combination of methodologies in the study of the same phenomenon" (cited in Jick, 1979, p. 602). Robson (2002) points to several advantages of using a multi-method approach, including a reduction in errors and its efficiency in tackling different research questions and a variety of subjects. Although Robson is aware of the potential theoretical criticism that social scientists might face in combining methods that belong to different theoretical traditions, he maintains that this integration could enrich the efficacy of social research in representing the object of investigation. On the same note, Jick (1979, p. 602) states "... researchers can improve the accuracy of their judgments by collecting different kinds of data bearing on the same phenomenon".

One of the methods employed in the current research was the case study method. Cases studies are widely used in social science to achieve an in-depth understanding of a unit of analysis (whether this is an individual, a group, an organisation or a geographical entity, etc.) (Robson, 2002). Researching the specific characteristics of a case study provides the opportunity to expand scientific knowledge by deepening the understating of how a wider social phenomenon is altered by such specific characteristics. The case study method also offers insight into the contextual conditions in which the case is embedded (Yin, 1981, 2003). There are, however, some disadvantages to the case study method. Certainly, the adoption of this method is going to limit the generalisability of the results. In fact, results might be generated by the specific characteristics of the context of the case study. For example, the outcome of a local debate about a proposed wind farm might be influenced by the type of co-operative scheme proposed—the feature of scientific interest— but also by other specific characteristics of the community, such as average incomes and education in the area or the presence of a local tourism industry, to name a few examples. Yin points out that case study research does not aim to generate a theory generalisable for the whole population, but rather to "generalize findings to the theory analogous to the way a scientist generalizes from experimental results to theory" (Yin, 2003, p. 38). In this way, a theory evolves through subsequent contributions based on empirical studies and is held as externally valid once a considerable amount of empirical research has confirmed it or, put another way, it has not been falsified (Popper, 1990).

I chose to limit my attention to case studies in the UK for practical (time and financial resources) and theoretical reasons. In the UK, the cultural trope of the "rural idyll" (Woods, 2005) might mean more resistance to any significant human modification of the rural landscape. This would be even more likely to be the case if, as Woods (2005, p. 4) notes, the rural idyll contributed to myths surrounding the construction of the UK's national identity, which is linked with the countryside. In this respect, the UK is a particularly interesting location for researching the different variables that influence the acceptability of wind farms and community energy schemes.

First study

The Westmill Wind Farm (in Watchfield, Oxfordshire) was chosen for the first case study. At the time of the literature review, it was the UK's first wind farm co-operative that sought full ownership by the local community.

The Westmill initiative stirred a considerable amount of local debate, but at the data collection stage, the share offer was not yet open. Hence, local residents were awaiting a later stage of the debate that would provide them with further information about the project and, in particular, its financial aspects. People involved with the project had developed established views on the subject and could also easily report on the perceived support that the project was attracting as well as on local views of contextually significant benefits and costs associated with the wind farm. In fact, some of the research participants promoted public initiatives aimed at locals in support of or opposition to the project; some were elected local government members who were sensitive to their constituents' views. Furthermore, the vast majority of respondents resided in the area, which is comprised of a few villages, and were thus likely to have a sense of local opinion towards the wind farm proposal.

Finally, stakeholders were the best informants on the topic of environmental citizenship. Being themselves at the forefront of the debate, or as local authorities, they were in an optimal position to answer questions regarding environmental citizenship and the relation between citizenship and the environment more generally.

Potential participants were identified through a report produced by the Thames Valley Energy Agency (TV Energy, 2004). Once identified, potential participants were contacted by telephone and invited to participate in interviews. The resultant group of volunteers was composed of a fairly equal number of supporters and opponents and included local authority and parish council representatives. Fourteen respondents were surveyed, 11 of whom were interviewed face to face, while the remaining 3, who were not available for face-to-face interviews, answered a questionnaire which included the same open questions that were posed to all the other interviewees. The data was therefore collected through semi-structured interviews. Appendix B presents the interview guide along with is rationale.

Second study

The second study was informed by the first study and by an updated literature review. The second stage of data collection was intended to survey local residents who lived close to proposed wind farms in order to assess how the co-operative model could influence the acceptability of the proposed wind developments.

A quantitative study would attribute a quantitative measure to various factors that were believed to shape the acceptability of wind farms, and particularly of wind farm co-operatives.

Because the highest volume of British wind resources are in Scotland, the research focused on four proposed wind farms in that country. In 2017, Scotland already had an approximately 59 per cent share of the total megawatt (MW) capacity of UK onshore wind farms (ONS, 2018). Crucial to the research was collecting data at the right stages of wind farm development. Certainly, the critical moment in terms of social acceptance for a wind farm is during the planning period, when the proposal is brought to the attention of the local population. As discussed in Chapter 2, research has shown that opposition is stronger before the construction of wind farms and that this weakens after construction (Warren *et al.*, 2005). Moreover, the period before planning permission is granted is the only period during which opponents have a chance of influencing the planning process.

In order to compare the co-operative model with commercial schemes in both high-income and low-income areas, the proposed wind farms of Bracco in North Lanarkshire, Meikle Carewe in Aberdeenshire, Cushnie in Aberdeenshire and Nigg Hill in the Highland Council were chosen. Income was a subject of interest because, as previous scholarship has illustrated, this is a personal resource that can alter a person's perception of the benefits of a proposed local wind farm and therefore, possibly, of a community wind farm co-operative.

Project size was also of interest. Because size is a factor that tends to make wind farms controversial in the local area, I was interested in looking at cases that were comparable to current commercial developments.

Given these conditions, only a few proposed wind farm co-operatives could be found on the Energy4All[1] website (a UK co-operative that promotes wind farm co-operative schemes in the UK) that met these criteria. At the time that the cases were selected, the only projects bearing an element of co-operative community ownership were Cushnie, Dunbeath and Nigg Hill wind farms. This meant that they were going to be only partially and not fully community owned through a co-operative scheme. Considering the difficulty in realising a commercial-sized development that is also completely owned by the local community, it was unsurprising that no such project was being proposed in Scotland.

The difficulty in gathering the funds to finance a wind farm of considerable size locally was an issue for the Westmill project surveyed in the first study. During the share offer, the project received a considerable amount of funding from The Midcounties Co-operative (formerly Oxford, Swindon & Gloucester Co-op), which bought 75,000 shares (Westmill Co-operative, 2006a) and is its largest shareholder. The business can be considered local, but it cannot be considered part of the community of local residents. Further, the share offer prospectus did not present a geographical limitation on investors. Only in the event that the sale of shares exceeded £3.75 million would investors within 50 miles of the wind farm have preferential access to the shares (Westmill Co-operative, 2005). This condition was effectively achieved, but as it was announced that the sale of shares achieved "over £4 million" it was stated that some applications would have to be "scaled back", implying that a vast number of applicants who did not live within 50 miles of the site would have had their applications accepted (Westmill Co-operative, 2006b).

The difficulty of raising local funds for locally-owned wind farm co-operatives has also been highlighted by a report on the financing of wind farm community schemes (TLT Solicitors, 2007). The report considers this to be a weakness of the co-operative model. However, partial community ownership through a co-operative scheme allows local residents to join and benefit from the scheme and to feel a sense of proprietorship.

It seemed reasonable to select two of the three schemes earlier mentioned (specifically, the most and least affluent) in order to facilitate comparison. The proposed installed capacity of the four Scottish cases needed to be as similar in size as possible in order to have comparable wind farm proposals, because the literature reviewed suggested that size would have influenced support and opposition. Dunbeath was therefore ruled out because of its far larger proposed size, which was expected to be 23 turbines and an installed capacity of 69 MW (West Coast Energy, 2005). At Cushnie, on the other hand, seven wind turbines and a total installed capacity up to 21 MW were planned (Falck Renewables, 2008), while Nigg Hill was expected to have five wind turbines and an installed capacity up to 12.5 MW (Falck Renewables, 2010).

In seeking commercial developments for comparison, it was found that there was a much larger population from which to choose. Using the Scottish Index of Multiple Deprivation (SIMD) interactive map[2] provided by the Scottish Government, and RenewableUK's online map of wind farms in planning, developments were chosen in the most income affluent and the most income deprived areas of Scotland, respectively.

The proposed Bracco wind farm in North Lanarkshire was identified as the development in the most income deprived area, and it was seeking consent to establish a power plant of seven turbines with a total generating capacity of 21 MW.

Identifying a proposed wind farm development in an income affluent area was more difficult. Most proposed wind farms are in rural areas, while Scotland's most affluent areas are located in its cities. Eventually, the Meikle Carewe wind farm in Aberdeenshire, a proposed development with 12 wind turbines and a total generating capacity of approximately 10 MW, was chosen.

Table 3.1 summarises the proposed wind farm cases. Along with the technical and geographical data regarding the proposed wind farms, i.e. the number of turbines, total MW of capacity, the local authority to which they belonged and the area's average urban/rural classification by the Scottish Government, it also presents social data regarding the levels of income deprivation and multiple deprivation.

Table 3.1 Cases chosen for the quantitative postal survey

Wind farm	Local authority	Turbines	Total MW	Type of scheme	Average % of income deprived	Average SIMD rank	Average urban/rural classification
Meikle Carewe	Aberdeenshire	12	10	Commercial	5.00	5414.78	2.91
Cushnie	Aberdeenshire	7	21	Co-op	6.15	4972.03	6
Nigg Hill	Highland	5	10	Co-op	15.86	2892.39	5.16
Bracco	North Lanarkshire	7	21	Commercial	18.09	2162.58	1.82

In order to assess the level of deprivation of the cases to be researched, 2008 data on the percentage of the population that is income deprived[3] at data zone level, supplied by the Scottish Neighbourhood Statistics Programme[4] (SNSP), was used.

Once selected, each data zone's population estimates for the year 2008, made available by the SNSP, were used to compute the percentage of the population that was income deprived in 2008. Table 3.1 presents the average percentages of the population that were income deprived in the areas surveyed around the wind farms.

For the purpose of better illustrating the social context of the areas surveyed, the average 2009 SIMD[5] rank of the areas surveyed was also calculated. To do so, both the rank values of the SIMD 2009 and the population estimate values of 2008 (made available by the SNSP[6]) were computed at data zone level.

The SIMD is a rank that indicates whether an area, A, is more or less deprived than area B. However, if, for example, area B has a rank value of 4000, it cannot be assumed that its level of deprivation is half that of area A with a rank value of 2000. Nevertheless, averaging the SIMD ranks of the data zones will generally indicate whether the area surveyed around Bracco is more deprived than the area surveyed around Meikle Carewe. The results of these 2009 SIMD rank averages are presented in Table 3.1.

Both the percentage of the population that is income deprived and the SIMD 2009 ranks concur, i.e. they present the same hierarchy of deprivation, ranging from the least deprived, Meikle Carewe, through to Cushnie, Nigg Hill and finally to Bracco, the most deprived.

Despite efforts to ensure relative parity among the wind farm developments surveyed, it is important to note that the Bracco wind farm is unique in that its nearby settlements are close to two large cities: Glasgow and Edinburgh. This location makes Bracco's surrounding environs only partially rural; indeed, most of the area surrounding the proposed wind farm is urban or semi-urban. This is confirmed by the urban rural classification[7] provided by the Scottish Executive for the Scottish data zones. The area surveyed around Bracco's proposed site has a mean urban rural classification of 1.82 with just a few data zones classified five, "accessible rural", and six, "remote rural". For the area of Cushnie, the mean is six; thus, all of the data zones within the area surveyed are classified as "remote rural". For Nigg Hill, the average is 5.16, which again means there is a high incidence of rural areas; and for Meikle Carewe, an area that has an almost equal number of urban and rural areas, the average is 2.91.

Pilot study

A pilot study of the survey was conducted, which targeted the case of Aikengall II, close to Dunbar in Scotland. Aikengall II was surveyed because it was a case in a relatively affluent area which could have been part of our final group of case studies. Eventually, it was determined that Aikengall II would not be included because it was too close to a pre-existing wind farm and the proposed development was too large in comparison to the other cases that were examined.

Eighty households were selected randomly within two bands located 0–5 km and 5–10 km from the designated site, respectively. Various authors have highlighted the role of distance in shaping attitudes about wind farms (Braunholtz and McWhannel, 2003b; Warren *et al.*, 2005; Bishop and Miller, 2007), and at least two studies (Braunholtz and McWhannel, 2003b; Warren *et al.*, 2005) have indicated that 0–5 km and 5–10 km bands of distance are useful partitions of the local area for research purposes.

The mailed questionnaire consisted of four pages printed on the front and back—a total of eight pages of text. It included several ranking questions that asked respondents to rank the two or three items of most importance within a larger group of several items.

The initial questionnaire was mailed with a letter. After two weeks, a reminder letter was sent that offered a second questionnaire on request if the first had been misplaced.

A response rate of 17.5 per cent was achieved. From the returned questionnaires, it appeared that some of the ranking questions were often not answered or were answered incorrectly (e.g. choosing one item instead of ranking items). This problem eventually led us to eliminate ranking questions from the questionnaire and to make almost all the items 5-point Likert scale questions. The final questionnaire is presented in Appendix A and is further discussed later in this chapter.

Sample determination

The formula proposed by Cochran (1977) for continuous variables that are investigated in a survey design is:

$$No = [(t)^2 * (S)^2]/(d)^2$$

where No equals the sample number of subjects to be selected; t is the t value corresponding to the probability of committing a type I error (the null hypothesis is rejected, but is correct [also called α, alpha level]); S stands for the standard deviation of the main variable under consideration; and finally, d is the acceptable level of error for the mean of the variable being estimated.

This formula was amended because the study had a two-group design, affluent/deprived, co-op scheme/commercial scheme, which should have been reflected in the formula of sample determination. Furthermore, to increase the accuracy of the sample determination, the "power", $1-\beta$, which is the probability of avoiding a type II error, β, beta (i.e. the null hypothesis is accepted, but is wrong), was included in the formula. In this study sample, calculation power was conventionally set at the level of 80 per cent.

Therefore, if the true difference in mean opinion is 0.5, then the number given by the formula would provide an 80 per cent chance (power) of detecting a statistically significant difference between groups at the 5 per cent level.

$$No = [(t + z)^2 * 2 * (S)^2]/(d)^2$$

with $t = 1.96$, $z = 0.84$, the standard deviation of the variable "opinion on the wind farm" obtained in the pilot study is $S = 1.157$ and the acceptable level of error $d = 0.5$ then $No = 84$, which is the number needed of each group co-operative/commercial and affluent/deprived. Thus, the total sample size is $No*2 = 168$.

A response rate of 20 per cent was considered achievable in light of the 17.5 per cent response rate to the pilot questionnaire, which was not only two pages longer and more complex, but had also been mailed out over the Christmas period.

Therefore, to achieve a 20 per cent return rate of 168 questionnaires, the questionnaire would need to be mailed to 840 households. As a precaution, 1000 questionnaires were also sent by email, 250 for each of the chosen wind farm areas.

Questionnaire development of the quantitative study

When designing the questionnaire, the research questions presented earlier were addressed by testing the findings of the qualitative study and surveying issues that previous research on the social acceptability of wind farms has deemed relevant.

A full version of the questionnaire is presented in Appendix A. There were minor wording variations across the different cases under consideration, depending on whether the questionnaire was delivered to residents living close to a proposed commercial wind farm or close to a wind farm co-operative. The layout of the questionnaire included in Appendix A was modified slightly to accommodate numbering of the questions, which was not included in the original version.

*Questionnaire administration, data quality checks
and response rate*

The questionnaire was administered through the postal service after randomly choosing 250 addresses for each site located in the two distance bands of 0–5 km and 6–10 km from the wind farm sites.[8] Overall, a total of 1000 households were selected. Following recommended research practice (Edwards *et al.*, 2002), several steps were adopted in order to increase the response rate. The selected households received a first letter along with a questionnaire and a prepaid stamped envelope. The letter introduced the research purpose and stressed the national importance of public opinion to the energy policy debate. After a week, a reminder letter was sent, and after another week, a third and final letter with another copy of the questionnaire and an additional prepaid stamped envelope was mailed.

The questionnaire was kept as short as possible in order to achieve a better response rate, in keeping with Edwards *et al.*'s (2002) determination that questionnaire length was the fourth most important characteristic in determining response rate. The questionnaire was limited to three pages, printed on both sides, and used Arial font size 10 for readability's sake, as suggested by Bradburn *et al.* (2004b).

Data from returned questionnaires was entered in SPSS (Statistical Package for the Social Sciences). Data quality was later checked by randomly selecting 50 questionnaires in order to cover the entire questionnaire set uniformly. When a mistake was identified in the questionnaire's entered variables, adjacent questionnaires (cases in SPSS) were also checked. Each questionnaire had 51 variables; over 2500 variables were checked and only 6 mistakes were found, i.e. the error rate was 0.24 per cent.

The questionnaire had a 31.5 per cent response rate: 315 questionnaires returned of 1000 mailed. The response rate in itself was an improvement over the pilot questionnaire's 17.5 per cent response rate and compares favourably with other postal survey research on onshore wind farms, many of which had lower response rates—e.g. only 13.3 per cent in the case of Swofford and Slattery, (2010).

Notes

1 Available at: https://energy4all.co.uk.
2 Available at: http://simd.scotland.gov.uk/map.
3 The percentage of "population income deprived" was calculated using data for the following categories of citizens:
  • number of adults (aged 16–59) receiving Income Support (Department for Work and Pensions [DWP], April 2005)
  • number of adults (aged 60 plus) receiving Guaranteed Pension Credit (DWP, May 2005)
  • number of children (aged 0–15) dependent on a recipient of Income Support (DWP, April 2005)
  • number of adults receiving (all) Job Seekers Allowance (DWP, April 2005)
  • number of children (aged 0–15) dependent on a recipient of Job Seekers Allowance (all) (DWP, April 2005)
  • number of adults and children in Tax Credit families on low incomes (Her Majesty's Revenue & Customs [HMRC], August 2006).

For further details see: Office of the Chief Statistician (2009).

4 Data available from: www.sns.gov.uk/default.aspx [Accessed 19 October 2010].
5 From www.sns.gov.uk/ [Accessed 25 October 2010]:

> The Scottish Index of Multiple Deprivation (SIMD) provides a relative ranking of the data zones in Scotland from 1 (most deprived) to 6505 (least deprived) based on a weighted combination of data in the domains of Current Income, Housing, Health, Education, Skills and Training, Employment and Geographic Access and Crime (SIMD 2006 onwards).

6 Data available at: www.sns.gov.uk/default.aspx [Accessed 19 October 2010].
7 The Scottish Government Urban Rural Classification provides a standard definition of rural areas in Scotland. It distinguishes between urban, rural and remote areas, based on a core definition of rurality which defines settlements of 3000 or fewer people as rural, and aims to assist understanding of the issues facing these areas. The Scottish Government has used the classification to improve the rural evidence base. It distinguishes between the following six categories: (1) large urban areas; (2) other urban areas; (3) accessible small towns; (4) remote small towns; (5) accessible rural areas; and (6) remote rural areas. For further information see the Scottish Government website: www2.gov.scot/urbanrural.
8 Addresses were identified using the Royal Mail's Postal Address file.

References

Bamberg, S. and Möser, G. 2007. Twenty years after Hines, Hungerford, and Tomera: A new meta-analysis of psycho-social determinants of pro-environmental behaviour. *Journal of Environmental Psychology*, 27, 14–25.

Bartlett, J. E., Kotrlik, J. W. and Higgins, C. C. 2001. Organizational research: Determining appropriate sample size in survey research. *Information Technology, Learning, and Performance Journal*, 19, 43–50.

Bauwens, T. and Devine-Wright, P. 2018. Positive energies? An empirical study of community energy participation and attitudes to renewable energy. *Energy Policy*, 118, 612–625.

Bishop, I. D. and Miller, D. R. 2007. Visual assessment of off-shore wind turbines: The influence of distance, contrast, movement and social variables. *Renewable Energy*, 32, 814–831.

Bradburn, N. M., Sudman, S. and Wansink, B. 2004a. *Asking questions*. San Francisco, CA: Jossey-Bass.

Bradburn, N. M., Sudman, S. and Wansink, B. 2004b. *Asking questions: The definitive guide to questionnaire design* (revised edn.). San Francisco, CA: Jossey-Bass.

Braunholtz, S. and McWhannel, F. 2003a. *Public attitudes to windfarms. A survey of local residents in Scotland*. Scottish Executive, Edinburgh. Available at: www. scotland.gov.uk/socialresearch.

Braunholtz, S. and McWhannel, F. 2003b. *Public attitudes to windfarms: A survey of local residents in Scotland*. MORI Scotland, Edinburgh. Available at: www.scotland. gov.uk/socialresearch.

Bryman, A. 2004. *Social research methods*. Oxford: Oxford University Press.

Cochran, W. G. 1977. *Sampling techniques*. New York: John Wiley & Sons.

Corbetta, P. 1999. *Metodologia e tecniche della ricerca sociale*. Bologna: il Mulino.

Corbetta, P. 2003. *Social research: Theory, methods and techniques*. London: Sage Publications.

Denzin, N. K. 1978. *The research act* (2nd edn.). New York: McGraw-Hill.

Diekmann, A. and Preisendörfer, P. 2003. Green and greenback: The behavioral effects of environmental attitudes in low-cost and high-cost situations. *Rationality and Society*, 15, 441–472.

Dobson, A. 2003. *Citizenship and the environment*. Oxford: Oxford University Press.

DTI. 2004. *Co-operative energy: Lessons from Denmark and Sweden*. Department of Trade and Industry, London.

Edwards, P., Roberts, I., Clarke, M., Diguiseppi, C., Pratap, S., Wentz, R. and Kwan, I. 2002. Increasing response rates to postal questionnaires: Systematic review. *The BMJ*, 324, 1183.

Falck Renewables. 2008. *Cushnie wind farm non-technical summary*. Available at: www.cushniewindenergy.co.uk/Userfiles/File%5CCushnie%20NTS%20PDF.pdf.

Falck Renewables. 2010. *Nigg wind farm non-technical summary*. Available at: www.niggwindenergy.co.uk/Userfiles/File%5CApplication%20Documents%5C2010%20SEI%5C2010%20Non%20technical%20summary%5CNon-Technical%20Summary_Low%20Res_v1.pdf.

Gillham, B. 2000. *Developing a questionnaire*. London: Continuum.

Groothuis, P. A., Groothuis, J. D. and Whitehead, J. C. 2008. Green vs. green: Measuring the compensation required to site electrical generation windmills in a viewshed. *Energy Policy*, 36, 1545–1550.

Henwood, K. and Pidgeon, N. 1994. Beyond the qualitative paradigm: A framework for introducing diversity within qualitative psychology. *Journal of Community & Applied Social Psychology*, 4, 225–238.

Hirst, P. 1994. *Associative democracy: New forms of economic and social governance*. Cambridge: Polity Press.

Hirst, P. 2002. Renewing democracy through associations. *The Political Quarterly*, 73, 409–421.

Inglehart, R. 1977. *The silent revolution: Changing values and political styles among Western publics*. Princeton, NJ: Princeton University Press.

Jick, T. D. 1979. Mixing qualitative and quantitative methods: Triangulation in action. *Administrative Science Quarterly*, 24, 602–611.

Klick, H. and Smith, E. R. A. N. 2010. Public understanding of and support for wind power in the United States. *Renewable Energy*, 35, 1585–1591.

Maslow, A. H. 1987. *Motivation and personality*. New York: Harper & Row.

Office of the Chief Statistician. 2009. *Summary of methodological changes to the Scottish Index of Multiple Deprivation (SIMD)*. Office of the Chief Statistician, Edinburgh. Available at: www.scotland.gov.uk/Resource/Doc/933/0097285.doc.

Oishi, S., Diener, E. F., Lucas, R. E. and Suh, E. M. 1999. Cross-cultural variations in predictors of life satisfaction: Perspectives from needs and values. *Personality and Social Psychology Bulletin*, 25, 980–990.

ONS. 2018. *Regional statistics 2003–2017: Installed capacity*. Office for National Statistics, London. Available at: https://assets.publishing.service.gov.uk/government/uploads/system/uploads/attachment_data/file/743827/Installed_capacity_of_sites_generating_electricity_from_renewable_sources__2003-2017.xls?_ga=2.131848271.1175064298.1566562378-907690731.1566562378.

Oppenheim, A. N. 1998. *Questionnaire design, interviewing and attitude measurement*. London: Continuum.

Popper, K. R. 1990. *The logic of scientific discovery*. Crows Nest, New South Wales, Australia: Unwin Hyman.

Rifkin, J. 2002. *The hydrogen economy: The creation of the worldwide energy web and the redistribution of power on earth*. Cambridge: Polity Press.

Robson, C. 2002. *Real world research*. Oxford: Blackwell Publishers.

Scottish Government. 2010. *Scottish Planning Policy*. The Scottish Government. Available at: www.scotland.gov.uk/Resource/Doc/300760/0093908.pdf [Accessed 10 March 2010].

Sheldon, K. M., Elliot, A. J., Kim, Y. and Kasser, T. 2001. What is satisfying about satisfying events? Testing ten candidates psychological needs. *Journal of Personality and Social Psychology*, 80, 325–339.

Shen, J. and Saijo, T. 2008. Reexamining the relations between socio-demographic characteristics and individual environmental concern: Evidence from Shanghai data. *Journal of Environmental Psychology*, 28, 42–50.

Swofford, J. and Slattery, M. 2010. Public attitudes of wind energy in Texas: Local communities in close proximity to wind farms and their effect on decision-making. *Energy Policy*, 38, 2508–2519.

TLT Solicitors. 2007. *Bankable models which enable local community wind farm ownership: A report for the Renewables Advisory Board and DTI*. Department of Trade and Industry, London. Available at: http://webarchive.nationalarchives.gov.uk/+/www.berr.gov.uk/files/file38707.pdf.

Toke, D. 2002. Wind power in UK and Denmark: Can rational choice help explain different outcomes? *Environmental Politics*, 11, 83–100.

TV Energy. 2004. *TV Energy case study: Westmill Wind Farm, Oxfordshire*. Available at: www.tvenergy.org/pdfs/westmil-web-case-study.pdf.

Wahba, M. A. and Bridwell, L. G. 1976. Maslow reconsidered: A review of research on the need hierarchy theory. *Organizational behavior and human performance*, 15, 212–240.

Walker, G. 2007. *Community energy initiatives: Embedding sustainable technology at a local level. Full research report*. ESRC End of Award Report, RES-338-25-0010-A. Economic and Social Research Council, Swindon, UK.

Wallace, D. 1954. A case for—and against—mail questionnaires. *Public Opinion Quarterly* 18, 40–52.

Warren, C. R. and McFadyen, M. 2010. Does community ownership affect public attitudes to wind energy? A case study from south-west Scotland. *Land Use Policy*, 27, 204–213.

Warren, C., Lumsden, C., O'Dowd, S. and Birnie, R. 2005. "Green on Green": Public perceptions of wind power in Scotland and Ireland. *Journal of Environmental Planning and Management*, 48, 853–875.

West Coast Energy. 2005. *Dunbeath Wind Farm environmental statement: Volume 4: Non-technical summary*. West Coast Energy, Maes Gwern Mold, Wales. Available at: www.falckrenewables.com/~/media/Files/F/Falck-Renewables-Bm2012/pdfs/our-business/elenco/Non%20technical%20summary%20for%20the%20project.pdf.

Westmill Co-operative. 2005. Westmill Wind Farm co-operative share prospectus.

Westmill Co-operative. 2006a. *09/05/2006 Westmill board announcement*. Available at: www.westmill.coop/westmill_newsdetails.asp?newsID=14.

Westmill Co-operative. 2006b. *Westmill Wind Farm co-op blown away by share success*. Available at: www.westmill.coop/westmill_newsdetails.asp?newsID=15 [Accessed 1 March 2006].

Wilson, N. and McClean, S. I. 1994. *Questionnaire design: A practical introduction*. University of Ulster.

Woods, M. 2005. *Contesting rurality: Politics in the British countryside*. Aldershot, UK: Ashgate.

Yin, R. K. 1981. The case study crisis: Some answers. *Administrative Science Quarterly*, 26, 58–65.

Yin, R. K. 2003. *Case study research*. Thousand Oaks: CA: Sage Publications.

4 The first community-owned co-operative in the UK

Lessons from Westmill Wind Farm

The following sections present the analysis of the qualitative data collected through semi-structured interviews conducted in the case of Westmill Wind Farm Co-operative, which formed the first study conducted in the research project presented in this book. The data are presented in several thematic units that recall the concepts earlier introduced in Chapters 1 and 2. As is often the case, qualitative research has the merit of not only being a means of possibly corroborating hypotheses, but also of bringing to the attention of researchers new variables that were not initially considered in the literature review. This was the case in this qualitative study, which allowed me to refine the choice of variables later used to inform the quantitative survey whose results are discussed in Chapter 5.

In the excerpts that follow, "I" indicates the author-interviewer, while "R" means respondent. All names have been omitted to retain anonymity.

Citizenship, environmental citizenship, responsibilities and rights

Citizenship

The majority of respondents, when asked about the concept of citizenship (question 2, Appendix B), demonstrated a loose understanding of the concept (as shown in Excerpt 1, quoted below). Some dismissed the question as too philosophical or irrelevant, and only one made an explicit link with the subject of the environment. An example of the understanding of the concept was provided by the leader of an environmental charity that supported the project. His response is below:

Excerpt 1
I: *What is the meaning that you attribute to that word?* [Citizenship]
R: The meaning, ehm, citizenship means the role that you play in society. You know, what society can do for you and your responsibilities within a community.

Often, respondents identified both of the two main dimensions of rights and duties pertaining to citizen status (Reeve in McLean and McMillan, 2009), although the theme of responsibility was more frequently mentioned. This is reflected in the answer of a local Parish council head:

Excerpt 2

I: *... or of your Council, in particular, I would be happy for you to highlight this because it could be useful for me. So, being a councillor, of course, you deal with the concept of citizenship on a daily basis, what does it mean this concept to you? The concept of citizenship.*

R: In relation to what we try to do in the village, I suppose, my own view is one of almost self-governance ... that in a village situation, you quite often get people saying they have seen someone dropping litter or they've seen someone doing something wrong and they come to the Parish Council and they say "you ought to do something about it" and, you know, what we say is, "well actually, you ought to do something about it because you are as much a citizen of the village as any others" and so it is a little bit of encouragement to self-governance really.

Citizenship at different geographical levels and conflict

Citizenship was largely defined as membership of a national community or society in general. Belonging to an international community was mentioned in only two cases, while the local community was named in just one case. Although this detail may seem minor, it may evince different conceptions of citizenship with conflicting priorities. Different conceptions of citizenship would motivate substantially different behaviours such as, for example, support or opposition to the locally proposed wind farm.

Dobson (2003) elucidates the difference between environmental citizenship and ecological citizenship, conceptualising a dual level of the spatial foundation for these memberships. With "environmental citizenship", the nation-state is the reference unit that defines the rights and duties towards the environment, while "ecological citizenship" is defined as non-territorial. In respect of the latter, Dobson refers to a form of citizenship that exists outside of the formal public sphere and which also encompasses the private sphere (Dobson, 2003, p. 39). Ecological citizenship is beyond the formal sphere and is a matter of subjective perception, the feeling of belonging to global society, or to the planet more widely. This emerges clearly from several interviews in which respondents made explicit reference to a sense of responsibility originating from belonging to the planet, or global society, although they usually did not link this to their stated concept of citizenship (see excerpts 3, 4 and 5 below).

A member of the wind farm opponents' group explains his sense of responsibility towards the global environment in the following excerpt:

Excerpt 3

I: *So we are talking about carbon dioxide and the production is the result of human activities, within our society, do you think that we are all equally responsible or is there some sector of society that is more responsible? What do you think about responsibility within society in general, is it an equal responsibility or not?*

R: Well, we all share planet Earth, our base of living so, in that sense, we all have responsibility to do what we can to keep it in good condition, and certainly, we have serious responsibility not to harm our environment in a wilful manner … Perhaps those are strong words. In some parts of the world, an impoverished human community will engage in what is called slash and burn practice, destroying forests in order to create some lands on which they can grow crops for their food. I understand that their conditions of life are very much impoverished compared to mine and I hesitate to blame them, but of course they are engaged in a practice which is harmful to our environment, and if we can find some way to avoid them doing this, I think that would be good.

Below are the words of a former parish and district councillor:

Excerpt 4

I: *In your views what is or who is responsible for climate change, especially in our national society? Is there any sector of society that you think is more responsible or is the responsibility widespread across all citizens. What is your view?*

R: I think obviously the government has got to take a lead in any sort of moves that are going to take place because, after all, the government is there to represent the people and should be capable of having a longer view of the world and our environment and our society than, you know, the ordinary citizen in the street. I think the government has the main lead responsibility, but I think following on from that there are other stakeholders. I think business is an important stakeholder because they have got to behave responsibly and they have got to develop processes that are environmentally sensible and at the very least don't do any harm to the environment, even if they don't help it. And I think as individuals we all have a responsibility because we all live [on] the planet and I think that is a responsibility for each of us to do what we can, you know, to help the planet, if not to become more safe, than at least to become no more dangerous than it currently is.

The interviewee in Excerpt 5 was a member of a renewable energy agency that supported the wind farm.

Excerpt 5

I: *You mentioned environmental values, but at the same time it seems to me that you were kind of mentioning other values as well. Can you say something more about other values that are close to the environmental ones? If you meant some other values as well …*

R: Sure I sort of mixed it up really but, well for me, ehm, part of, well this is all my personal opinion … and it's … responsibility to the environment is part of the same ideology, if you like, as responsibility to fellow people, to fellow animals and it's all wrapped up in the same idea. That's why I was using it sort of interchangeably almost and some people delineate responsibility to humans and responsibility to the rest of existence ...

I: *You talk of combining together these ...*

R: Yeah, yeah, that's it. And you can rationalise that by saying we all depend on the environment and the integrity of the environment is crucial to our survival, conflict etc. But for me, those are extra arguments rather than the fundamental argument which is that we have a responsibility to the world as well as having the right to survive and be comfortable.

Environmental citizenship, the sense of belonging to the national community, was not directly mentioned as a source of perceived responsibility. This might be due to the fact that the first question asked respondents about climate change, which is rightly considered as a global phenomenon, and may have led respondents to indicate membership of the global community as a source of responsibility.

Nevertheless, several cases hinted at environmental citizenship but did not explicitly mention it. For example, in Excerpt 6 below the words "our communities" may refer to the national community, or perhaps to the world community as a whole. The person interviewed here was a former parish and district councillor.

Excerpt 6

I: *You mean ... you are talking about the moral commitment of people then?*

R: Yes, yes

I: *It is quite interesting, because I was about to ask you, to introduce a specific question regarding rights and duties in relation to the environment, and so probably you were starting already to answer this question ... What is your opinion about rights and duties in relation to the environment?*

R: Again, I think we have certain duties to the community around us which means that we don't pollute, we don't destroy the environment around us because it doesn't exist just for us as individuals, it exists for all of our communities as a whole. I think it's difficult really to go beyond that because an awful lot of what individuals do can be important in very small ways but not in very big ways, whereas what some small group of individuals who have a lot of power—economic power—do can be very disruptive.

Yet the statement in Excerpt 6 ("I think we have certain duties to the community around us ...") could also be seen as underlining the sense of belonging to the local community.

The sense of belonging to the local community, a kind of perceived "local citizenship", appears as an important theme motivating actions against the wind farm (see Excerpt 7 below). In the next passage, the respondent, a member of the opposition group, clearly contrasts the concerns expressed by the leader of the supporters' group and the wind farm applicant—regarding the threat that climate change poses for communities in the developing world—with his concerns for the local community. He uses expressions of emotional belonging such as "loving your neighbours". His sense of belonging to the local community thus served as the main motivation for his opposition.

Excerpt 7

R: He [the wind farm applicant] said to me, "Why don't you care about the children in Bangladesh that are going to be drowned?" I said, "[name omitted] look around you ..."

I: *Why don't you care about ...?*

R: The children in Bangladesh that are going to be drowned as a result of climate change, that's the sort of thing that he will use to make people feel bad, ehm and I said to him, what about loving your neighbours, the persons who live near you, and the impact that you have on them and what you are doing? And that to me is what matters more, as you can see, I mean it was acknowledged that there will be, I think, over 200 people in this community who will be badly affected by the low-frequency, medically badly affected, why should that be allowed to happen? Just because everybody says it's better than climate change, so yeah I do have ... I feel our community has been attacked by this and is not going to be healed by being offered shares in something ... that's not even a sticking plaster.

Both the proposer of the wind farm and a member of the supporters' group plainly conceptualised the conflict between local and ecological citizenship in Excerpts 8 and 9, respectively.

Excerpt 8

I: *Yes, well, what is your opinion about local community involvement in protecting the environment? There are advocates of local community involvement that say that a local community can provide a very good knowledge of the environment and so good environmental solutions, while critics say that local communities could be involved but there would be the problem that they could be conditioned by their local interests, so sometimes prevent good environmental solutions. What is your opinion about that?*

R: Well, I guess that in relation to wind farms that's quite complicated because the ... when you say look after the local environment, then the local environment is, you know, visual, aural. The wind farm is looking after the wider environment [but] is looking after the wider environment in terms of renewable energy so, I think, there is clear potential for conflicts of interest between what may be perceived as looking after the local environment and resisting industrialisation or development within an environment, and looking after the wider ... environment and promoting renewable energy schemes. I can see there is a conflict of interest there. So as a local individual I can understand that people would want to resist a development if they did not believe or understand the bigger implications of it because they could see it as actually damaging their local environment.

I: *Mmm ...*

R: Does this answer your question?

I: *Yes, yes, it makes sense, mmm... I think it's quite controversial this matter of local community involvement because in some cases the local community can bring an environmental project alive, contributing actively and fostering it, and in other cases they can prevent it being realised because they feel that it's a threat to their lives in some way ...*

R: A threat to their lives (and goods) in some way?

I: *Yes, to their environment in some way so ...*

R: Well, definitely.

I: *It's quite controversial ...*

R: Yeah, I mean nuclear power, I mean I think you know another example is nuclear power stations, locally, would be a threat to the local environment but nationally or globally in terms of CO_2 reduction of emissions, it could be seen as a very positive development ...

I: *Yes, that's true*

R: So I think it's a conflict between the local environment and the wider environment ...

Excerpt 9

I: *What is your opinion about citizens' rights and duties in relation to the environment?*

R: I guess, as a citizen, I suppose I see citizenship, as I think you are using the word ... as within, you know, more of either a locally defined citizen of your area or within a nation-state, and therefore I suppose once the relationship to [your] environment is ... Yes, we have responsibility to our fellow citizens for a local environment ... but as, you know, a member of the global citizenship, yes, obviously, you have wide responsibilities which are wider than just within my local environment.

Warren *et al.* (2005, p. 854). designate the conflict of interests between local residents that support a wind farm in order to benefit the global environment, and those who oppose it in order to maintain the integrity of the local environment, the "green on green debate". But while the authors are concerned with the conflict between environmentalists, in the case of Westmill Wind Farm it extends beyond those who would label themselves as environmentalists to those who wish to benefit either the local or the global community.

Local citizenship, place attachment and place identity

"Local citizenship" is related to the concept of "place attachment" (PA). Local citizenship is intended here in the sense of belonging to the local community, and PA is "defined as the affective relation or the emotional bonds that people have with places where they live" (Bonaiuto *et al.*, 2002, p. 636). However, in Excerpt 10, the respondent (a member of the opposition group) shows an affective relation to the local environment that connects to the sense of belonging to the local community.

Excerpt 10

I: *Well, I understand, well as a farmer and you said before you probably have clear ideas about what it's important to do in order to defend the environment and the importance of living in a clean environment. So do you have any ideas about what are the rights and duties of people in relation to the environment?*

R: Ehm ... well our position on this farm has always been leave the land that we have been stewards of a little better than we took it on and that's my approach to life, whether it's towards people or whether it's towards the land, so ... Obviously we need to nourish it and look after it ... but we need to do that in a holistic way, in a way that encompasses everything and I don't believe it's happening with this ... this scarring on the landscape that is going to occur and the impact on the next generation, you see. The generation that grows up in Shrivenham now is going to grow up expecting that turbines on the landscape is an acceptable thing to do. What's that taught them about looking after their environment? It hasn't done, it's told [them] that it's OK to do that and I think that is very poor education and I feel my position ...

I: *Because you think that they are going to spoil the environment?*

R: Yes, obviously they are going to, because they are going to add that huge visual impact where you can't help ... I mean [name omitted] has put a tiny turbine down over there, whatever he is doing with that I don't know, it's spinning all day and when you are out on the farm all you look at is that tiny [turbine], probably it's only 6 feet across, but it draws your eye because it's moving, and do you see the trees around? Do you see the birds around? No, you see this spinning turbine. Now you put these massive great things up, and that's what this landscape is going to be now, as it was put in the EIA [environmental impact assessment], a "wind farm landscape", whatever that means. We all know, because I've been to see these big ones and I know what it does, not only visually but the noise, it's the whole thing, it's like saying, "it doesn't matter what impact we have on this landscape", and it does matter for people's hearts and minds, it does matter ... and in their own attitude to their own landscape.

In Excerpt 10, the respondent makes clear his belief that the turbines will substantially change the local environment and the perception of it that future generations will hold. Bonaiuto *et al.* (2002, p. 636) state: "Place identity was defined as that part of people's personal identity which is based on or built upon the physical and symbolic features of the places in which people live." A changed physical environment, therefore, could alter the place identity (PI). The respondent in Excerpt 10 speaks about the "poor education" future generations will have and the fact that they will view the turbines as a normal part of the landscape. The same cannot be said for the respondent himself; he views the turbines as "scarring on the landscape" and made it clear that this "scarring" would impact the affective relation that locals have with their environment—their place attachment.

However, are PA and PI the source of a negative attitude towards the turbines or the consequence of the same independent variable that generates such an attitude? Examining the creation of two protected natural areas in Italy, Bonaiuto *et al.* (2002) suggest that PA and PI might be sources of opposition. In some ways, the creation of protected natural areas is similar to the construction of wind farms, particularly when their potential negative economic impacts are considered.

Bonaiuto *et al.* (2002) affirm that local communities negatively viewed the protected natural areas because they feared the change in land use would cause economic suffering: those who would enjoy the biggest advantages were non-locals who could visit the areas or simply enjoy the overall reduction of pollution. Similarly, in terms of reducing greenhouse gas emissions, wind turbines benefit the wider environment more than the local environment. In fact, local people might bear the costs, such as the visual impact, and some fear additional consequences, such as noise, declining property values and possibly more. Hence, it might be that, ultimately, these perceived costs could move some residents, particularly those who live closer to the planned wind farm site, to protest. As a consequence of their action, they would develop a stronger PA. Further research is needed to clarify whether PA and PI are a consequence or a source of negative attitudes towards locally proposed wind turbines.

In the last decade, the role of PA in relation to social acceptance of energy projects has received considerable attention in research (Devine-Wright, 2009; Devine-Wright and Howes, 2010) although without achieving consistent results (Devine-Wright and Wiersma, 2020). Some authors (van Veelen and Haggett, 2016) have argued that place attachment can also play a role in motivating the establishment of community energy projects. Outside of research on local energy developments, some scholars (Kyle *et al.*, 2004; Xu and Zhang, 2016) have found that motivation, intended as "the pursuit of personal benefits" (Xu and Zhang, 2016, p. 89) appears to be one of the antecedents of PA. This therefore suggests the possibility that, in the case of wind farms, motivations related to the perception of the benefits and costs due to the siting of a proposed wind farm might be an antecedent of PA.

The co-operative model

Based on the work of Toke (2002) and Diekmann and Preisendörfer (2003), among other authors, it was hypothesised that a locally-owned co-operative scheme would gain wider acceptance of the proposed wind farm because it would endow locals with both a sense of ownership and a financial incentive to support local wind farm development.

It was presumed that the financial incentive would compensate for the wind farm's visual impact. After reviewing the interview transcripts, however, these assumptions were revised.

First, creating a community-owned co-operative was more easily visualised than made a reality, as wind farm developments are often controversial. This was certainly the case in respect of Westmill Wind Farm. Supporters and opponents created two separate local networks and both were active in trying to persuade the local authority in charge of the planning permission process, Vale of White Horse District Council, and in encouraging others to take their side.

As participants and witnesses recounted, the local debate was bitter. One former district councillor reports:

Excerpt 11

> I was involved with the wind farm at Watchfield particularly because on the day after my election, ten years ago, having celebrated my success the night before, I had a ring at my door bell at 8.30 in the morning from somebody who wanted to know what my attitude was to the wind farm [laughs]. I [had heard] nothing about the wind farm at that point, having just been elected, I hadn't been briefed on any of the issues ...

The head of a local parish council, meanwhile, described the division as follows:

Excerpt 12

> Well, it has polarised the village and to the extent that other expressions [have been] made, half are for it and half are against it, now ... well no, I estimate half are against it, half are concerned ...

A member of a third-party organisation that backed the support group offered some insight:

Excerpt 13

I: *You were aware about this wind farm project in Watchfield?*
R: We were probably contacted initially by the farmer [name omitted] probably coming across by now, he wanted basically our support because of all sorts of very locally strong anti-contingency [events organised by the opposition group], he was then looking to get involved with environmental groups who see the broader global scale and aren't as interested in, you know, "not in my back yard" sorts of issues, so he contacted us [but we] haven't got involved on that specifically ...

Both the opposition and the support group acknowledged the division. A member of the support group explained why the debate over the wind farm had become so heated:

Excerpt 14

> The reason why we had such a fight on our hands, I am sure you know it already, is because people are frightened, and a lot of misinformation was made ...

And a member of the opposition group interpreted the conflict as follows:

Excerpt 15

> This community was a well-balanced, caring community with neighbours who look after each other and [take] care of each other, what more is community than that? And this idea has just caused complete divisions … It's gone crazy, it's, it's a type of dominance just remerging in a different form of people wishing to dominate other people under the guise of "green", and that's terrible.

It was not possible to establish whether local residents supported or opposed the wind farm. The survey commissioned by the applicant (Weston, 2002) did not directly measure support; rather, it asked respondents about their perception of the changes in the landscape. A slim majority of the total sample (51.5 per cent) expressed the opinion that the landscape would become "more unpleasant" or "more offensive" as a result of the construction of the wind farm (see Table 4.1).

The conflict between the support group and the opposition group is likely to have prevented high participation in the co-operative among residents of villages in the wind farm's immediate vicinity, although no detailed data was available to confirm this assumption. The share offer was extended throughout the UK and large investors were admitted; thus, the previous limit of 80 shares per person was abandoned (Midcounties Co-operative[1] bought 75,000 shares). In this respect, the Weston (2002) survey cited above offers some insight into low participation in the co-operative at the local level, as 59.5 per cent of the sample said that they would not have invested in the co-operative.

One specific factor exacerbated the division: Westmill was not initially conceived as a community-owned co-operative, but rather as a single farmer's personal initiative. The subsequent push towards a co-operative, possibly community-owned, was, in turn, perceived by some community members—and particularly by opponents—as a mere ploy to gain the community's support and thus increase the chance of success in the planning process.

This is how the personalisation of the project was perceived by a member of the supporters' group:

Excerpt 16

I: *Why do you think that was not understood, the sense of a co-operative?*

R: I mean we campaigned and said, you know, it's going to be the local community that will benefit but I don't think that people really understood that they could actually put their money in, even though we said you only need £250, you know, to invest. I just don't think people realised that it actually could be theirs and I think that partly [was due to] the fact that the original application went in 12 years ago and it was a fairly small wind farm and it was seen entirely as [the farmer's] baby and he was going to profit, so I think it took a long … I don't think we ever got over that idea that he was going to be the beneficiary, even though he consistently said "this is …" you know, "I want to do this because I want people to be involved".

Table 4.1 Expected landscape attributes following the construction of Westmill Wind Farm

	%		%		%		%		%		%
More interesting	61	No change	36.5	No change	53.5	More active	52.5	More managed	40	No change	54.5
No change	16.5	More unpleasant	32.5	More enclosed	17.5	More busy	23.5	More industrial	22.5	More unsettling	31
More bland	10	More offensive	19	More exposed	12.5	No change	20	No change	18.5	More threatening	9
More boring	6	More pleasant	10	More open	10	More tranquil	2.5	More commercial	18	More safe	5
More invigorating	6.5	More beautiful	2	More confined	6.5	More peaceful	1.5	More natural	1	More comfortable	0.5
	100		100		100		100		100		100

Source: Weston, 2002.

The co-op promoter and landowner make it clear that this personalisation played a role, and actually admitted that he did not consider a community-owned co-operative model in the first application:

Excerpt 17
I: *Probably there is matter of communication as well, of course, you probably made all the efforts to try to give another kind of image of the project ...*
R: But if I said, even though, even from the beginning, for this planning application it was seen to be some community ownership and to be fair that wasn't there in the past, but for this one it was there, people have to believe what you say and didn't actually believe you. And why should they believe you? You know you haven't done that before, you have no experience, you have no track record of having done that, so they wouldn't necessarily believe you anyway. So, if they are open to believe you, they will ... just because I've said that ... doesn't mean that people, reasonably enough, have to accept it, because things can change. You know, maybe I've got planning permission, so I ... actually don't do it as community ownership ... you know, you know ... thank you very much. You know until that happens you don't know.

Several different sources confirmed that the landowner, in fact, had a leading role throughout the entire process of promoting the wind farm.

A member of the supporters' group explains how he came to join the group:

Excerpt 18

... I became involved in that because I taught environmental science to my students and they needed to know about alternative energy as part of the course, so I got in touch with [name omitted] because I knew that he was trying to get planning permission and he talked to the students and I got to know him a little bit through that and then bumped into him a few more times. From that he invited me, when he realised that it was just taking so long to get the application through, and he knew that I was very keen on alternative energy, he suggested that I came and joined the group, so I did. It was a very ad hoc group; we just chatted, you know, people came and went. It wasn't a signed-up membership or anything like that, and [name omitted] asked me sometimes to do interviews for the press or to contact the press and just to be involved, so I was there really more as a support and giving ideas as to how we could progress and put pressure on and writing letters to the councillors and writing letters to the press and so on. I'd say that there was a lot of that kind of activity, that it was done very much with the group, you know, and different people helped at different times.

The active promotion of a network of supporters by the wind farm's landowner also emerged from the transcripts, supporting the idea that he was always seen as the prominent figure promoting the wind farm. In Excerpt 13 we saw a pertinent account from the head of a local environmental group that backed the wind farm of how a network of supporters was mobilised by the farmer.

In the context of a personalised project, the co-operative scheme incurred the risk of being perceived as a move to attract public consensus, rather than as a genuine move towards community energy ownership. In fact, the opposition exploited this argument, as the below interview with a member of the opposition group demonstrates:

Excerpt 19
I: *Well, do you think that the fact of selling the shares in the local community is going to make any difference in terms of the opinion that people might have about the wind farm?*
R: I just think that it's a clever ploy to try to cover up the divisions that occurred in the community. It's trying to buy people off and trying to tell them that they are going to make something out of it and that they are going to feel better in themselves by doing it, and I think it's cynical, I do.

And again, from the same interviewee:

Excerpt 20
I: *They basically claim that it has got an educational value and a revenue for the local community, because of the shares ...*
R: But that's just buying them off, isn't it? That's just buying off the community to stop the community from actually saying "No, we don't like them, no".

Interestingly enough, a member of a non-profit organisation that backed the wind farm project echoed concerns that the co-operative scheme and its revenue could be perceived as an attempt to buy the community's consent:

Excerpt 21

... in terms of specifically, ehm, community benefits, then that can have great impact in the planning stage. Particularly, it can swing a community for or against it depending on whether they perceive it as bribes, or they perceive it as fair contribution to the community to compensate for the visual disamenity.

The perception that the initiative was ultimately promoted by a single person may have made some suspicious that its advantages would primarily accrue to the co-operative's founder. It is worth noting, moreover, that the wind farm's key promoter was a politically active semi-public figure who was probably appreciated by some locals and despised by others.

One member of the opposition group expressed a view concerning the suspected personal advantage for the landowner:

Excerpt 22

Well, to set up any kind of renewable energy scheme is going to cost money and presumably it would be unreasonable to expect anybody to do it without some kind of recompense for so doing. On one occasion, at a meeting [name omitted], the applicant here, was accused of only being interested in the money and his bland reply was "well, we all need money!" [laughs]. Which is true, but it does lead you to question slightly his real commitment to the environment.

Community owned or potentially divisive?

Thus far, the uniqueness of this case and how it might have affected the local debate has been highlighted. Still, it can be contended that some degree of opposition would have been present regardless of whether the project was perceived as a personal or a community initiative.

Opposition to a wind farm proposal, depending on its ability to involve locals, could create a rift in the community that prevents a large number of people from supporting a co-operative scheme. Hence, the result could be a co-operative that is partially locally owned but not strictly community owned, as a fairly large amount of people would abstain from investing.

This situation could also, paradoxically, lead to an established internal divide within the community, with the wind turbines being a constant reminder of the conflict. This point is expressed by the head of a local parish council:

Excerpt 23

I: *If we could compare renewable energy development owned by the local community and renewable energy developments that are not owned by the local community, what do you think would be the difference in terms of impact on the local community? Do you think that they could impact differently on the local community?*

R: I don't think so, I don't think so. I think the only difference, perhaps, might be that if you have a background of the kind of "acrimony" that we had here between the supporters and the opponents, I think if you have continuing local ownership amongst those people who were great supporters then it's just going to keep an open wound festering, and I think that's the downside to it ... They would just stop the wound from healing because these people will be going to their meetings and we'll be having, you know, the publications going out, details will be going out, and people who were against the wind farm in the first place ... just will not like that and there will be a constant reminder of a battle lost.

Another ex-councillor expressed his view about the potential for dividing the community:

Excerpt 24

... this polarisation thing, "local co-operative" implies that everybody invests in it and we know that everybody won't, so once again, you get a polarisation of those that are in the club and those that are outside the club.

And, finally, the promoter of the wind farm also mentioned this issue:

Excerpt 25

> Are there any disadvantages? [...] Well I suppose, as the potential to divide the
> community between people who think that's a good thing or a bad thing and
> therefore some people making money out of it and others choose not to invest in
> it, and then there's an element of, you know, potential for some sort of division.

Just two interviewees—a member of the supporters' group and a representative
of a supportive organisation—were more optimistic that once the wind farm was
built, local residents would accept it after realising that its impact was not
negative.

Compensation

Compensation for the visual and other impacts of the wind farm is only available to people who ultimately invest in the scheme. Although the co-operative
could consider offering some free services to the local community (at the time of
data collection, it was not known whether it would), it is likely that these would
be a minor compensation in comparison to the financial rewards that would
accrue for co-operative members.

A survey conducted by Oxford Brookes University (Weston, 2002) revealed
that the vast majority of local residents (75 per cent) believed that the co-operative
should provide some sort of community benefits, though nearly 60 per cent of the
sample said that they would not invest in it. Community members thus want to
enjoy the advantages of a co-operative without making any sort of financial outlay.

The wind farm co-operative promoter and landowner highlighted the limits of
the compensation that a co-operative scheme could provide:

Excerpt 26

> Yeah, well I think that if we own something we are obviously more well-
> disposed, you know, benignly disposed towards it ... so if you own some-
> thing, there is the balance between the financial benefit that you get from
> that and the inconvenience, whatever that may be. The situation with
> community-owned projects is that only a certain proportion of the com-
> munity will subscribe to them, so it doesn't really address that [opposition
> towards proposed wind farms], but it goes some way towards it. I guess the
> other way to address that is for the project to commit a proportion of its
> funds as income towards what would be seen as a local organisation repre-
> senting the community, be that the parish council or the school or the Scouts
> or whatever. Maybe that's the other way or [providing cheap] electricity which
> is very complicated, but ... it would be good if it does that and certainly that
> is what this project's hoping to do, both through local ownership and then

providing a source of income to the local community ... Yeah, yeah, but I don't think community ownership hits it [squarely] on the head because some people won't have any ownership of it and they will be looking on, hearing, or thinking they hear, or whatever might be, or have the distress of having a development near them, if they feel distressed by that. And for people who feel genuinely distressed, they [don't want to] own it anyway so for the people who are most [affected] that's actually a poor compensation ...

The possibility that the co-operative's revenue might be a compensation was also strongly dismissed by the chair of a local parish council:

Excerpt 27
I: *Well, advocates of this community-owned co-operative scheme say that, for example, for wind turbines, the revenue coming back from the scheme to the local community could compensate for visual impact and noise. What do you think about that?*
R: You can never "compensate" for visual impact. There's nothing you can do to compensate for that ... And as for the other issues, I think they are largely more about fear than actuality, but once again, it's how you get people that are worried to a stage where they cease to be worried, and therefore I don't think it's about the money side of it if you like ...

A local councillor suggested the possibility that the project was less attractive and thus investment was lower in such an affluent area:

Excerpt 28

... I think it makes sense in certain contexts like the sort of examples that we were talking about earlier, you know, Scotland and so on, but I think, in area like this which is a relatively affluent middle-class area where people have got expensive houses, there is really nothing that you can offer them in terms of economic benefits, so to that extent what they would see is these benefits. But in other places, I think where, you know, there is a lot of local unemployment where you have got the older industry like mining and steel or what have you, which have died and people are unemployed, I think [that's] got to be a part of the selling point there for them, but it's very much a situational thing.

Although the councillor's point about the area's affluence is understandable, in the case of a community-owned co-operative, the opposite could be true. As socio-economic status (SES) models of civic participation suggest (Verba *et al.*, 1995), affluent individuals are more likely to participate in society and, in this specific case, wealthy people could more easily afford to buy shares in the co-operative. Regarding this, it is worth noting that in the Oxford Brookes University survey (Weston, 2002) cited earlier, one respondent admitted that she could not afford to buy any shares.

Overall support could be higher in an economically-depressed area if it was believed that the co-operative would generate jobs but, paradoxically, this higher level of support for the co-operative would be unlikely to be expressed through buying shares.

The fact that local people's appreciation for wind turbines depends on the characteristics of their location and whether they are interested in exploiting the economic opportunities created by the wind farm was suggested by a member of a public organisation committed to promoting renewable energy that backed the Westmill Wind Farm Co-operative project. This is, moreover, consistent with the literature on visual perception of turbines in different environments (Lothian, 2008; Molnarova *et al.*, 2012).

Excerpt 29

> ... I am sure that you have heard about the two turbines there, those Ecotricity [a UK renewable energy supplier] turbines. [The] first one was part of the environment centre, Ecotech, ehm, this is a place in Norfolk, they put the first one with a viewing platform, and it's a particularly nice part of the country. I should say I am from that area and we have a landscape that's suited to tall objects because it's very flat like Holland and so people are open to it, and they supported the turbine. Even the opponents reported that once it was up, they could go up and view from the top of the turbine and it's becoming a massive tourist attraction for the town and the local community demanded a second one in the company's consultation and so now they have built a bigger one. That's also in the town and so it's an example of what can be done and what a sense of ownership can do even without the economic ownership.

While the limits of the co-operative model in terms of overcoming opposition have been highlighted thus far, some respondents considered the community-owned co-operative scheme to be superior to purely commercial developments, provided the wind farm was situated far away from the village in a rural area. A member of the opposition group made this clear:

Excerpt 30

> ... in theory, I could envisage a community such as ours which had the opportunity to support a wind farm that was on a hilltop 2 miles away on wild [land] which would not have a serious impact upon the community. I hope people would leap at the chance of supporting it and of doing something, a little step if you like to renewable energy sources.

This respondent's strong pro-environmental attitude based on accurate knowledge of the current debate about climate change and energy sources was evident throughout the interview.

The reason why the co-operative scheme had a better chance of gaining local support than a privately-owned outside company was because the profits would remain within the community. A local councillor who opposed the wind farm commented on this:

Excerpt 31

> ... in theory if the community generated the project then they'd be more likely to support it, if a big company corporation did it then probably it wouldn't get a lot of support, because it would be seen as being a purely profit-making thing, you know ...

A member of a partner co-operative that had utilised a similar scheme to Westmill (partially locally owned) explained the scheme's ability to influence undecided local residents:

Excerpt 32
I: *How do you think that local ownership of wind farms influences levels of opposition or support for wind farms?*
R: It gives a voice for supporters and help sways [those who are] undecided but it's not going to affect strong antis who are a minority 5 per cent of the population and vocal.

The original promoter of the wind farm believed that the scheme could potentially attract more support than a private company—even though that did not happen in this case, mainly because of the inability to effectively communicate the potential benefits for the community. Another unfavourable factor that he underlined was the fact that people might be unaware of similar successful schemes because they were new to the UK; awareness might play a role in facilitating acceptance of a community-owned co-operative wind farm:

Excerpt 33
I: *Ok, about levels of opposition and support for the wind farm, in particular this one, but in general all wind farms, do you think that local ownership can make a difference?*
R: I don't think it has, I think it should do, I'm surprised that it hasn't. I suppose it is because people don't understand what it is really. It's to [engage] in a concept that most people don't quite get their heads around. So, it hasn't made very much difference but it can do if the issue is clearly explained and what the benefits are, and I think if there are more examples [that were] happening, yes, it will make more difference in the future.

He continued to build his argument, mentioning the fact that the lack of established and successful community-owned co-operative schemes made people more suspicious of the proposal and less likely to join:

Excerpt 34
R: ... if they were you know, relatively neutral—would they have been less neutral if there was [co-operative] ownership? It may make a difference ...

Yeah, I think in the future it will make more difference than at the moment when there are more examples of projects like this, I think. And then other communities can say why can't we have some of the [action] too because [they've got it there]. There are almost no examples of that and they aren't publicised, so when a wind farm planning application is proposed the opposition [say] "let's stop it!" … not let's have some ownership of it.

A member of the opposition group also pointed out that co-operative community-owned schemes were unknown in the UK, when asked about factors other than economic incentives that could encourage people to join the co-operative.

Excerpt 35
I: *Do you think that there are other factors that influence people in buying a share, apart from economic revenue?*
R: I don't think that I am in a position to offer an opinion on this, there is no comparable situation anywhere else in Britain; whether there is in continental Europe, you will probably know better than I do.
I: *There is a case in Britain as well, the Baywind Co-operative in Cumbria.*
R: That's right, yes, but this is not purely local ownership. The term "local" is a relative one; they are gathering funds from people originally in the north of England, in Lancashire particularly.

Are wind farms a "high-cost situation"?

The theoretical framework outlined in Chapter 2, which combines rational choice and attitudinal theories, could help to explain the behaviours of support and opposition towards community wind farm co-operative developments. In doing so, it is necessary to speculate whether support for wind farms have high- or low-cost implications for local residents.

For many people, wind farms are likely to be perceived as having high-cost implications, particularly before the construction of the wind turbines when they can only imagine the impact that the turbines will have on their lives and risk overestimating some costs.

Concerns about noise, devaluation of property prices and visual impact may be the core factors that influence negative attitudes towards a proposed local wind development. Negative perceptions of noise and visual impact are over-emphasised by opposers who consider these factors to be real threats to the well-being of local communities.

In the case surveyed, anti-wind farm activists reported concerns about noise, particularly low-frequency noise; in general, local residents in the vicinity of proposed wind farms frequently mention noise as one of their main concerns (Warren *et al.*, 2005).

Some consider the wind farm's visual impact to be a serious issue, to the point that they anticipate feeling distressed by a landscape that they would consider unpleasant. As the survey commissioned by the proposer (Weston, 2002, pp. 11–12 and 33–35) indicated, the community around Westmill was worried about the aesthetics of the wind farm: several respondents expected that the local landscape would change after construction to become more unpleasant (32.5 per cent) or more offensive (19 per cent) with regard to the physical features, and more unsettling (31 per cent) and more threatening (9 per cent) regarding the sense of wellbeing that the landscape provided (see Table 4.1). Some respondents considered visual impact and noise to be health costs of the wind farm, and some believed both affected emotional wellbeing.

In terms of economic costs, the threat of property devaluation could motivate some to oppose the wind farm, as several interviewees made clear in their answers. Although research has suggested that property devaluation is non-existent in the case of wind farms (Sterzinger *et al.*, 2003), perception of this risk is what matters to local residents during the pre-construction phase. An additional economic cost that was not relevant to the specific case at hand, but which certainly applies elsewhere, is the perception of adverse effects on the local tourism industry. The Vale of the White Horse does not have a well-developed local tourism industry, thus it did not apply here.

On the other hand, what are the possible benefits that could accrue to the local community and counterbalance the costs mentioned above, and with the same level of importance? Chief among these is the economic revenue deriving from the wind farm that could be used to offer free services to local residents, or the distribution of a proportion of it within the local community. Job creation could be an additional hypothetical indirect economic benefit for the community (Halliday, 1993; Warren *et al.*, 2005) and could be particularly attractive in economically deprived areas, while it would have minimal appeal in affluent areas where the unemployment rate is low.

The Westmill co-operative scheme attempted to offer economic benefits to whoever was willing to invest in the wind farm which, in theory, might have appealed to the local community. Interviewees did not perceive the co-operative model's economic opportunities as being very strong; however, the interviews were conducted before the investment prospectus was circulated, and perceptions might have changed over time. During the local debate, the perception was that such economic benefits were comparable, if not less attractive, than others offered by the then current investment market and, further, some respondents questioned the financial viability of the wind farm. During the interviews, nobody expressed a desire for the wind farm to create jobs or provide free services to the community. Some respondents made clear that the local community was quite affluent, hence its members would not have been easily swayed by such economic benefits. Table 4.2 lists the costs and benefits cited during the interviews.

The cost perception is, of course, subjective, yet it appears to be related to physical proximity, as discussed in Chapter 2. The Oxford Brookes University survey of local residents living around the Westmill site shows that community members who lived close to it expressed greater concern about the visual and auditory impacts of the turbines than community members who lived further away.

As discussed earlier, the "inverse NIMBY" hypothesis (Warren *et al.*, 2005) contends that attitudes change before and after construction. People who live

Table 4.2 Costs and benefits identified in the case of Westmill Wind Farm Co-operative

Costs	Benefits
Health: • noise pollution • distress caused by visual impact	Economic: • attractive investment • free services provided to the local community
Economic: • devaluation of property prices	

closer to a wind farm site oppose the development more before its construction but, eventually, those same individuals become the most supportive once it is built and becomes operational.

How can this attitudinal shift be explained? A possible explanation is that opposition and support are expressed as the result of an evaluation of the expected costs and benefits associated with the proposed wind farm. The perception of those costs, however, is based on expectations; most respondents are likely to be unfamiliar with wind turbines, apart from indirectly, through media coverage that often concentrates on controversies and therefore stresses the costs for local communities that host wind farms. These expectations may be overly pessimistic as a result of the anxiety that respondents confronted with the perceived risk of some negative consequences might experience.

After construction, though, the costs and benefits are reassessed and, as a result, opposition decreases —or support may even develop. By this point, many more individuals consider the situation low-cost or even beneficial and, under the influence of positive attitudes towards renewable energy, end up being supportive of the wind farm. The same shift is less likely to happen in respect of those who live further away from the development, because the turbines are not part of their immediate surroundings and thus they are less likely to reassess the turbines' impact on their lives. They may, however, retain old prejudices related to possible negative features such as noise, property devaluation or visual impact. They may erroneously assume that these negative impacts are true for the area immediately surrounding the wind development but, because they do not live there, they are unlikely to have their assumptions challenged.

A second, complementary hypothesis that could explain local people's cost reassessment aligns with Rippetoe and Rogers' stress theory (1987, in Gardner and Stern, 2002). Rippetoe and Rogers (1987, p. 598) define avoidance as "an attempt to evade actively or deny the threat" as an emotion-focused coping strategy in response to a stressor. It could be hypothesised that some subjects force themselves not to be disturbed by the turbines once constructed in order to avoid any unpleasant feelings or stress that the opposite response would provoke in them.

The two hypotheses outlined above are not mutually exclusive: for example, an individual could perceive the turbines as much less noisy than they expected and deny any annoyance as a result of that noise level.

While it is not possible to provide evidence confirming the "inverse NIMBY" syndrome in Westmill during the post-construction phase, it is worth noting that the two leading figures in the opposition group lived closer to the proposed wind farm site than any other interviewees.

Costs and benefits in the Westmill case

Noise pollution, a health cost?

Both this study and the Oxford Brookes University survey revealed that noise was one of local residents' greatest concerns. In the Oxford Brookes survey (Weston, 2002, p. 37), 64 per cent of respondents said that the larger turbines proposed in the final planning application would be noisier (38 per cent) or significantly noisier (26 per cent) than the smaller turbines that had originally been approved. This means that local residents considered noise perceptible if they believed that they would be able to notice the difference between smaller and larger turbines.

Opponents often mentioned "low-frequency noise" as a particularly worrisome consequence of the wind farm and believed that it would negatively affect the health of the local community. One member of the opposition group showed great concern, having actively sought information regarding this topic:

Excerpt 36

> Recently, research by a Dutch man, a Dutch professor of Physics who is very highly regarded across the world in his knowledge of low-frequency vibration, finds that vibration which you cannot hear but which does impact on our bodies—we are talking about vibrations of less than once per second—one of the sources of these vibrations is the pressure created when the blade of the wind turbine comes past the column. It sends a little shiver through the column and that vibration passes through the ground for up to 10 kilometres. Now, there has been much research in America, particularly, but elsewhere also into the effect of low-frequency vibration on the human body. NASA, for example, are very concerned about low-frequency vibration impacting on helicopter cruises causing disorientation, loss of concentration, dizziness, headaches and actual blackouts to which some people are very much more susceptible than others; some people are virtually immune, it is accepted, but not everyone. Now, the health of a big community could be at stake. I would see it as absolutely proper that that community and its representatives do ask a lot of serious questions as to how much is really known about such phenomena, and what it is the likelihood of serious damage.

These worries were shared by another activist from the opposition group, who evoked strong images of serious harm to human health by providing a dramatic personal account of meeting some people affected by low-frequency noise because they lived close to a wind farm and describing the use of low-frequency noise as a weapon of war.

Excerpt 37

R: I talked to one woman and I went to see her in her home and it was … her husband said to me, "I don't need to know which way the wind is blowing in the mornings, I simply have to look at Gwen [the man's wife] to see how ill or well she is and then I know" that from their [the turbines'] position … and it was a southwest wind that blows up to the low-frequency … into their home and they had to move and you know …

I: *They had problems caused by the low-frequency?*

R: Yeah, yeah …

I: *So was not just the visual aspect …*

R: No, no, it's the low-frequency, which is something we have all campaigned against because there are people in Shrivenham that are part of the Army who have been trained about this in terms of weapons of war, using low-frequency as a weapon of warfare and the research shows that the trouble was that they couldn't. It was affecting their own soldiers, its impact was devastating, and so we knew that there was a problem potentially. It's only now, unfortunately for us, that there is now a report, actually commissioned by the DTI [Department of Trade and Industry] to show that there is no problem, that actually has proved there is, up to 10 kilometres, a very real problem of low-frequency that affects people … You know it will become a public health issue and we should still be going back to our planning office and saying, "Are you going to take this down? These are breaking the rules". We are not going to stop because we are not going to stop speaking about the fact that these things are harming us, if they don't, great, let's hope they don't.

Such alarmism is not confined to the opposition group: individuals who opposed the project but where not activists also cited concerns about noise as reasons for their opposition. A local councillor who was also a member of the planning committee expressed his concerns about low-frequency noise.

Excerpt 38

My main opposition to the wind farm was based on two [pieces of] medical evidence, which I saw which suggested that people living near the wind farm within 500 metres or so would be affected by low-frequency sound. This is one of the major things that [I've doubted], you know, I was a bit uneasy about, and these were papers which were written by doctors in the medical journals and these particular doctors live near to people who live near to a wind farm, do you understand? And they described various phenomena, one was Dr Amanda Barry who was actually going to come here to speak against the farm, you know at the planning meeting, but she … at the last moment, she couldn't come, you know, and I think that her evidence might have persuaded the planning committee a bit more. The main problem that she saw was that some people, not everybody … felt that particularly at night, they had a sort of a resonance effect, you know it's, it's like ... they couldn't hear it, you can't hear low-frequency sound, you know, below 10 Hertz it's impossible to hear, But they felt that there was something, sort of uneasy about themselves, you know, they were unable to sleep, they had what we call anxiety syndrome states ... and this might have been psychosomatic, you know, in their minds really, but we weren't convinced about the evidence, mainly because the patients, many of them described that when they went on holiday or even when they stayed with their relatives, there were no problems ...

Now, again, it wasn't [firmly] determined whether this was psychological or not, but then we had a professor here, who works at the military college of science, he is a military man and he confirmed that, in his opinion, low-frequency sounds could damage people and we then opposed it mainly on the basis that there was not enough work done, you know. There's plenty of work done on audible sound, in other words, what you can hear, but not enough work done on low-frequency sound and this was the only real matter that we ... I mean the visual effect was no problem, you know ... the fact that, OK, it was being built on a greenfield site which in this area would not normally be permitted, you know, if this was a turbine station, you know what I mean or a power station or anything like that we would not admit it, permit it. But because it was a green source of energy eeh ... it was considered to be OK for a greenfield site, you know, but the low-frequency sound was the main issue and ... This, I believe, is why the Danes and the Germans now tend not to build their wind farms close to a population area, because we've read some papers from a Danish work ... and German work as well, where the Danes were building much more offshore, you know, out to sea, [laughs] and the Germans have stopped actually building wind farms near to population areas, not only because of low-frequency sound but for other reasons, you know, so this was the reason why ... there was a small group of us on the planning committee who did not ... who were not very comfortable ...

Fears about low-frequency noise were also articulated by another councillor who opposed the project, although he also expressed his frustrations about how difficult he and the local community found it to obtain independent information about low-frequency noise and other issues related to the wind farm.

Excerpt 39

> ... and this is what has been difficult for us as a parish council—it has been very, very difficult to get information on the risks of the things, you know ... [With regard to] low velocity noise, it has been very, very difficult to get expert opinions because the people supporting the wind farm went on and got ... expert opinions, the ones that were worried about the wind farms went out and got another lot of expert opinions and they were miles apart ...

Nearly every respondent who opposed the wind farm cited noise and low-frequency noise as one of the main reasons for their opposition. There were a few exceptions. One organisation opposed the wind farm because of its statutory interest in the conservation of historical buildings, while an ex-councillor opposed the project because of its visual impact, claiming that the anti-wind farm campaign's use of the noise issue went to extremes that he considered absurd. Supporters knew that they would have to deal with the noise issue in order to rally support: one individual stressed how the opposition manipulated fears about noise and loss of property values to gain ground within the community.

In conclusion, the noise issue played an important role in determining opposition to the wind farm, while the specific issue of low-frequency noise was considered a potential threat to public health.

Distress originated by the wind farm's visual impact—a possible health cost?

The Vale of the White Horse is considered a pleasant, unspoiled rural English landscape. Apparently, there are no big visual intrusions on the western and eastern extremes that signal the limits of the valley. The wind farm site lies in a designated Area of High Landscape Value and is just 6 km away from an Area of Outstanding Natural Beauty (DPDS Consulting Group, 2002). The villages in the Vale are of historical interest and a 7,500 acre National Trust property neighbours the site. Finally, the Ridgeway National Trail passes not far from the site. It is therefore clear that the wind turbines, which would be up to 81 metres tall (DPDS Consulting Group, 2002), would have a visual impact of some concern.

All the interviewees who opposed the wind farm, including activists, local councillors and members of third-party organisations, mentioned their concern about visual impact and many also referred to the concerns of local residents. When local people (about 300 respondents) were asked how they thought the landscape would change after construction, replies showed mixed attitudes, with the majority of the sample expressing apparently contrasting views in respect of different items. For example, 67.5 per cent said that the landscape would be more interesting or more invigorating, while 51.5 per cent believed that it would become more unpleasant or more offensive and 40 per cent considered that it would become more unsettling or more threatening (Weston, 2002) (see Table 4.1).

Typically, a wind farm's visual impact is thought to motivate opposition because many local residents value the features of their landscape. However, in the interviews some respondents suggested that frequent exposure to a landscape composed of turbines could provoke real distress for local people. Literature on "stress theory" is consistent with this view of turbines as an environmental stressor: the theory defines an environmental stressor quite broadly as an object of one's environment that is perceived as unpleasant or seen as a threat to one's wellbeing to some extent (Gardner and Stern, 2002). The stressor is considered to affect individuals through two cognitive appraisals: a primary appraisal that recognises the stressor as such and a second one that establishes the subject's potential for control over it. If such control is impossible, the potential stressor becomes a source of stress and the subject will, in turn, develop coping strategies, such as focusing on pleasant features of their surrounding environment (Gardner and Stern, 2002).

In the following excerpt, a local councillor who opposed the wind farm acknowledges the link between visual impact and distress:

Excerpt 40

> I mean, if I had, ehm, if I bought shares in a wind farm, I think I'd [put up] with [laughs] the visual impact, you know, everything that is causing distress to others, [laughs] as long as it didn't affect me.

While a member of the opposition group avoided using the word "distress", he used similar language to describe the farm's negative visual impact:

Excerpt 41

> Its impact on what has been designated as an Area of High Landscape Value cannot be beneficial, and in the minds of many people could be seriously damaging. Unfortunately, personal reactions to a wind farm cannot be measured in terms of numbers; it is a subjective thing, and I'm aware that some people like wind turbines, comparing them with modern sculpture ...

The landowner mentioned distress in relation to the wind farm, although he did not mention visual impact as its source. However, he differentiated this distress from noise annoyance (see Extract 26 above).

In the Oxford Brookes University survey (Weston, 2002), 32.5 per cent and 19 per cent of the sample held the opinion that the landscape would become "more unpleasant" and "more offensive", respectively. In response to another question, 31 per cent thought that the landscape would become "more unsettling" and 9 per cent thought it would be "more threatening" (see Table 4.1). These results evince how a proposed wind farm may lead some subjects to anticipate a situation of distress.

Economic cost: the possible devaluation of property

Property devaluation constituted the only applicable economic cost of wind farm construction. Although many interviewees considered the Vale of the White Horse a place of particular amenity, nobody mentioned a local tourism industry that could have been negatively affected by the wind farm. Instead, three interviewees mentioned declining property values: a member of the opposition group; a member of the supporters' group; and an ex-councillor. It is interesting to note that, although other interviewees did not specifically mention property values, they nonetheless recognised it as a concern independently of their role in the local conflict surrounding the wind farm.

The following excerpts underscore the significance of the expected property devaluation issue to the opposition group, to the point that the former councillor suggested that it was their only real concern:

Excerpt 42

A lot of local opposition was orchestrated by local landowners. I think between the time [name omitted] put his first application in and the second and third applications, there was a very considerable change in the local environment, because one of the things that we find happening was that Swindon, which obviously is a major town just a few miles down the road from us, had decided that one of the things that it wanted to do was to expand eastwards, in our direction ... I think, particularly landowners who had thought that all they had was farm land, now potentially saw a few years down the road, the possibility that they might have very, very valuable pieces of real estate in their ownership and the sort of money that we are talking about is ... If you have agricultural land, at present around this area, it will probably bring in about £5000 an acre, [were] you to sell it; if the same land were to be sold as potentially development land, either for housing development or industrial development, it is probably worth in excess of £900,000. It may even be worth £1 million an acre. So when you start looking at the difference between £5000 and, you know, £1 million an acre, you are talking megabucks, and I think that people who previously felt that the wind farm perhaps didn't affect them too much, suddenly saw that, goodness, it was a real problem and it started to be an orchestrated campaign against the wind farm by the [name of the opposition group omitted] people in particular ...

And again:

Excerpt 43
I: *Yes, have you ever thought about any possible link between opposition and support of a wind farm and local ownership?*
R: No, I don't think local ownership in any way whatsoever affected the views that were taken here. I think there were economic issues coming from the opponents because they felt that the value of their land and the value of their properties and so forth was going to be affected because they were turning out the most outrageous horror stories. There was one guy talking from [name of the opposition group omitted] one time, I heard saying if you had a home within a mile of the wind farm then you wouldn't be able to sit out in your garden on a sunny summer's day without having ear protectors, or you would damage your hearing. You know, that's absolute lunacy, but these were the sort of things that were being said ...

None of the other interviewees linked opposition primarily to loss of property values. However, as mentioned above, a member of the supporters' group also admitted concern over the potential loss of property values:

Excerpt 44

> Yeah, the reason why we had such a fight on our hands, I am sure you know this already, is because people are frightened and a lot of misinformation was [given] and it's very, very easy to misinform. Ehm, and so people feared that it was going to be very loud, people ... some people genuinely don't like them, you know, visibly you can't change people's mind on that, but lots of hoo-ha was made about noise, which was just completely inaccurate and a lot of fear of loss of property value that's ... is there, I think it's in people's mind. Ehm, and I think ... some elements of [the] media have been very anti-wind farm and therefore anti this whole process, and I think those things have made it very much more difficult ...

Economic benefits: the revenue available to co-operative members and free services for the local community

The previous section discussed the suitability of the co-operative model for overcoming opposition and compensating for the wind farm's costs. Many of those interviewed expressed a negative opinion, but this only included people who were strongly against the wind farm. These individuals certainly viewed the possibility of joining the co-operative as "not even a sticking plaster", as one of the members of the opposition group put it. In essence, they believed that the wind farm's costs were going to be so great that no benefit could compensate for them.

In the Oxford Brookes University survey (Weston, 2002) a majority claimed that they would not invest in the co-operative scheme (about 60 per cent), suggesting that the local perception of costs was overwhelming the attractiveness of the community ownership scheme. This is striking considering that the area is wealthy, and hence a small investment would not have been a big sacrifice for most of the households.

However, for other respondents, provided they did not perceive the costs of having a local wind farm in their area as real, or were perhaps just slightly concerned about them, the possibility of buying shares and earning revenue was considered as a kind of benefit able to compensate or overcome the costs.

This was admitted by the opponents as well; this is shown in Excerpt 30, cited above, in which a local councillor admitted that some people may feel attracted to buying shares as long as they do not perceive the visual impact as affecting them. A member of the opposition group, stated that this economic benefit might work be seen as advantageous by a majority of the local community:

Excerpt 45

> In theory, I am sure that it is a very good idea if you can involve the local community by allowing them to buy shares and to profit financially from the operation of the ... wind farm. I think the whole community or a majority of it may be supportive and they will be prepared to accept the visual and possibly the audial impact ...

Shares were sold at £250 each, making them affordable to the vast majority of local residents, particularly considering that the 2001 Census showed that the local area is wealthy. In the statistical unit of Shrivenham (comprising the homonymous village and the villages of Watchfield, Longcot and Bourton), the unemployment rate was less than half the rate for England (1.48 per cent compared with an overall rate of 3.35 per cent) while educational level was higher, with 37.6 per cent of the population attaining level 4/5 (degree or higher degree) compared with 19.9 per cent for the population in England overall (ONS, 2001).

Opinions differed regarding the safety of the investment, with supporters claiming that it was a safe investment and opponents suggesting that investment in the co-operative would be a financial disaster, and claiming that supporters and the council had not been honest about the financial viability of the project.

Apart from enjoying the revenue, which was subject to buying shares, the scheme offered other possible economic benefits, such as free services to the local community. This possibility was mentioned briefly by the landowner and first promoter of the wind farm. The survey carried out by Oxford Brookes University (Weston, 2002) revealed that this possibility enjoyed broad support: 74 per cent of respondents said that the co-operative should provide community benefits. Given that such benefits are the only economic ones available to the entire community, they may offer the only compensation to people who oppose the wind farm and declined to purchase shares, as suggested by the first promoter of the wind farm (see Excerpt 26). Although respondents were clearly keen on receiving such benefits, they were uncertain as to what they were and the extent to which they would be delivered. Some may even have viewed such benefits as "bribes", as one anti-wind farm activist and a member of a third-party organisation which supported the wind farm indicated (see Excerpts 20 and 21). In sum, community benefits and open ownership were a rather weak draw in support of Westmill Wind Farm Co-operative.

Socio-economic status

Socio-economic status (SES) has been proven to positively influence political participation (Verba *et al.*, 1995). Participation in wind farm co-operatives is different: while the cost of being a political activist is mainly in terms of time, in the case of a co-operative, there is a financial cost as well. Furthermore, the physical presence of the wind farm will bring some costs to the local community, as earlier discussed, while a political process does not generate costs per se. Nobody in fact, in a democratic context, feels threatened by elections or political party meetings. Hence, the influence of SES might be very different in relation to wind farm co-operatives. On the one hand, high SES makes it easier for people to buy shares because income, education and probably a certain professional status might well make people more inclined to invest in a local wind farm—or, at least, prevent them from giving up because of lack of resources, both financial and in terms of ability to understand what it means to buy shares and invest in a co-operative scheme. On the other hand, local people, if affluent, may feel that they are not at all interested in the economic benefits that the wind farm might bring—in fact, they don't need them. This can be true for local jobs or local free services and, surely, for the revenue of the co-operative: they could use their money for other investments that may be even more profitable, even though may entail more financial risk.

A former local councillor (see Excerpt 29) expressed his opinion about the importance of SES in the Westmill case and how it actually limited the chances of support in an area such as Watchfield. The planning officer of a non-profit organisation who opposed the development expressed the same view:

Excerpt 46

I: *What about the economic revenue that people could get from the investment in shares of this local co-operative? Do you think that a possible economic revenue could compensate in some way the negative perception of the visual impact or noise ...*

R: I think that it could have a bit, yeah, some people who were prepared to invest the time to go into investigating the financial opportunities might, yeah, then, you know, that would influence their overall attitude if they felt that yes, there was ... a contribution to the local economy that could be substantial. Often, remote rural communities are suffering at the moment economically and we are talking of Westmill here, Westmill is in a very affluent area of the country, you know, Oxfordshire, every car, every house has got three cars and all of that but, well, not every house but, you know what I mean, it's a very affluent part of England ... Many other wind farm proposals are in much more remote landscapes in Scotland, Wales and Northern England and so on ...

I: *Even affluent people like to invest their money in profitable businesses.*

R: They do, yeah, they are very good at it [laughs] but what I meant was that the advantages to a local community, ... as a proportion, would be more significant in a rural community which is, you know, suffering because agriculture is suffering or ... there's not a lot of industry and not a lot of alternative sources of income.

Another councillor at both the parish and the district level, who opposed the wind farm, pointed to the fact that the most educated people also opposed the wind farm:

Excerpt 47

Peer influence is important because if, for instance, in Watchfield, where the wind farm was, if the majority of the opinion leaders over there would have supported it and would have invested in it, and you know, I think there would have been a much bigger uptake but that didn't happen. In fact, the more educated people were able to analyse that this possibly would not be such a good thing, you know, for various reasons, and they thought about that negatively and they influenced other people in a negative way. If the wind farm had been

... if proper figures had been produced, you know, on its viability and how efficient it would be and [what] the return would be and all the other things, if all the negative bits had been taken out and the opinion leaders over there had said, "Wow, this is good, we are going to invest in it, I recommend you do it, you know, we are going to make money, we are going to save the earth, we are going to do all the other things that ...", then I think the support would have been overwhelming, but unfortunately it turned out the other way ...

Pro-environmental attitudes and values: avoiding cognitive dissonance

An enduring strand of research asserts that pro-environmental behaviours are primarily influenced by values and attitudes (Stern *et al.*, 1999). This project takes a different approach, considering attitudes as a motivator that comes into play only when a low-cost situation is in place. This theoretical approach can explain why strong pro-environmental values, attitudes and even behaviours are expressed by individuals who fiercely oppose a wind farm.

Those interviewed expressed varying levels of environmental concern. All but two of them were concerned about climate change; and these did not deny climate change, but expressed some doubts.

Interestingly, both of the anti-wind farm activists interviewed exhibited strong pro-environmental values. One of them was an organic farmer who, during the interview, proudly displayed a biomass boiler that had recently been installed in a building on her farm that was to host a school for children with special needs:

Excerpt 48

> I think we all have a duty to use as little as we can, to recycle as much as we can, to ... like in this building with the sun pipes that we have used to design our building for maximum energy efficiency, we've put in a boiler so that we can use our own willow ... So it's not a matter of not putting our money where our mouth is ... or not caring at all, but we are caring in a way that's sustainable. These turbines are not ... the impact that that they have is ... it has a destructive effect. What we are doing here is not causing harm to anybody; ... we have put this building up [and] it's not affecting [name omitted] on the hill.

The organic farmer showed particular enthusiasm towards a specific kind of renewable energy, probably because, as a farmer, she was in the best position to value the potential of such a source:

Excerpt 49
I: *Well, there are other forms of renewable energy like biomass ...*
R: Yeah, that's so sensible, biomass is so sensible ... because it not only looks after the land; it gets jobs for the people and it does the job of creating energy and that's absolutely brilliant ...

Similarly, the other activist who opposed the wind farm showed great concern for the environment and climate change. He was conscious of the fact that society needs to take drastic action and that everybody must do their part. He also agreed that wind energy could be a useful part of the solution.

Excerpt 50

> I am a broadly-based scientist, I know perfectly well that the world stock of fossil fuel has been depleted very rapidly as a result of our use of fossil fuels. Greenhouses gases, so called, have been pumped into the atmosphere at an apparently increasing rate; this is causing climate change, global warming and "we are heading for disaster". Something needs to be done quickly. As a biologist, my feeling is that we are not doing anything like enough to adopt a policy of carbon dioxide capture. There is much talk about renewable energy, but very little about reining in the emissions of carbon dioxide and some other gases into the atmosphere, and the only solution, as I see it, is a massive change of lifestyle for the human community. This, I think, [needs to be] on a global scale really, although the civilised Western nations really have to give a lead in this respect. Do we really need to be forever going halfway around the world for a few days' holiday? In a jumbo jet? And, in many other ways we could, through adopting a change in lifestyle, put the brakes on this problem of global warming.

And about wind power, the opposition activist said:

Excerpt 51

> Unfortunately, persuading the rest of the world to change its way of living is not going to be easy, is not going to happen quickly, so, yes, we have got to go for renewable energy sources of which there are many. I do believe and accept that wind power in the right place is a very good response because it can be put in place fairly quickly. It is an expensive way of reducing CO_2 emissions, and I can give you figures for that, but the problem is serious and we have to bite the bullet and if it costs a lot of money well, so be it, we still have to do it.

It is fairly clear from the quotations of both activists that wind farm opponents held pro-environmental attitudes. Yet the perceived costs of the wind turbines outweighed these attitudes. These two activists, in particular, were neighbours of the wind farm site and they expressed concerns about the negative impact that the turbines might have upon them and the local community in general.

The organic farmer had a special reason for opposing the wind farm because she runs a school for children with autism at her property. She is a trained teacher of children with special needs and was concerned about the possible negative impact that the turbines might have on the children's sensory environment. Her willingness to host these children and to spend money to build this school on her farm (which was, as far as understood, just partially funded by a grant) demonstrates altruistic values often associated with pro-environmental behaviour (Stern *et al.*, 1999).

The Oxford Brookes University survey (Weston, 2002) found positive attitudes towards both renewable energy and wind power among local residents. It is therefore imaginable that in order to avoid cognitive dissonance with these prevalent beliefs, local people would have been well disposed to support the project; but considering the strong opposition and negative views that emerged in the survey, they may have objected to this specific wind farm. A possible explanation for this is that the perception of major costs ruled out any influence pro-environmental attitudes may have had in shaping the behaviours of some members of the local community, whose numbers are impossible to quantify.

Networks in competition: using communication to magnify costs or benefits

Wind farms are controversial because they usually generate conflict between proposers, their supporters and opponents. It is worth focussing a little more on the conflict that happened at Westmill and what it can tell us more generally about similar situations. How can a rational choice and attitudes framework help us to understand the dynamics of the conflict surrounding the wind farm siting and its purpose?

The Westmill survey transcripts reveal how social support was used as a weapon against opponents in the planning process, as suggested by one of the ex-councillors interviewed (see Excerpt 11). The planning process involves a commission of councillors—elected representatives of their constituencies. Hence, social support is a way of putting pressure on councillors because if they act against their constituents' wishes they risk alienating the public and the loss of future elections.

It is, therefore, useful to be able to mobilise people for or against the project and the way in which that seems to happen is through organised campaigns that magnify the costs or benefits of the project. The importance of public debate in shaping local attitudes is highlighted by Wolsink (2007), who contends that attitudes are formed through public debate.

In the case of Westmill, the first organised network created was in opposition to the wind farm. It was built by neighbours of the proposer who had received a letter from the council informing that the second planning application had passed; therefore, they gathered to discuss the matter. They later formed a group and started acting by recruiting consultants, writing letters to the council and approaching local politicians and newspapers. It is worth taking a close look at one of the flyers produced by this group. The following text is extracted from an original scan of the flyer. It was the first passage on the flyer (TV Energy, 2004):

> Did you know that a local landowner has applied for planning permission to erect 5 huge wind turbines on the north side of the A420 opposite

Watchfield and Shrivenham? If this development goes ahead our landscape will be ruined, our lives blighted by high levels of intrusive noise and the value of our properties will plummet. *If this catastrophe is to be prevented it is vital that YOU act now.* The danger is real and immediate [emphasis in the original].

This part of the flyer summarised all the points that were then developed in further detail in following sections; other arguments were introduced to support their stance too.

Stressing the costs that the wind turbines would have for the local community was the approach that appeared to be used to mobilise people; in doing so, these costs seemed to be emphasised. The flyer emphasised the costs listed, particularly those related to noise, visual impact and property devaluation, with the first two regarded as health costs (noise pollution and distress caused by the visual impact), while the third was considered to be an economic cost.

Both wind farm opponents and supporters organised actions in order to magnify either the costs (the opposition) or the benefits (the supporters) and hence mobilise people for or against the wind farm. Supporters organised public events that were aimed at providing information about the proposal, climate change and renewable energy.

In a weblog,[2] the wind farm promoter described the reasons for creating the group of supporters, saying that a response follows every action. The opposition group was a response to his planning application, while the support group was a response to the campaign of misinformation and scare stories. He continued:

Local people who had seen the [opposition group's] leaflets and TV coverage and knew that it was rubbish wanted to respond. A few phone calls and a bit of networking resulted in a meeting of nearly a dozen people around the kitchen table and so [the support group] was formed.

To respond to the negative campaign of the opposition group, the promoter organised the support group (in Excerpt 18, one member of the group states how she was recruited by the promoter). Both groups established a network of individuals and organisations that backed their actions. In particular, Friends of the Earth (FOE) and Thames Valley Energy got actively involved in campaigning in favour of the wind farm, while the Campaign to Protect Rural England (CPRE) actively supported the opposition group.

The support group, FOE and Thames Valley Energy, for example, organised a public exhibition, a public event called "Oxfordshire and Wiltshire Renewable Energy Festival" and a petition that was signed by over 1000 people. The opposition group organised public meetings, a petition and a wide distribution of flyers. Both groups approached the local newspapers and released press communications and interviews.

NIMBYism, an outdated concept to define opposition

NIMBYism (not in my back yard) has been widely used to describe the behaviour of local residents that oppose wind farms in their locale (Devine-Wright, 2005), although the term is also used to express local opposition towards a facility's siting (Wolsink, 2006).

Unfortunately, as Wolsink (2000, 2006, 2007) indicates, the NIMBY explanation for local opposition towards wind farms fails to provide an exhaustive account of a phenomenon that has much more complex motivations. A strictly NIMBY attitude towards local wind farms would lead individuals to reject local projects but to support wind farms elsewhere. Wolsink (2007, p. 1201) instead specifies four kinds of opposition to wind farms:

> I A positive attitude towards the application of wind power, combined with an intention to oppose the construction of any wind power scheme in one's own neighbourhood (the only true NIMBY-motivated opposition).

> II The not-in-any backyard variant, which means opposition to the application of wind power in the neighbourhood because the technology of wind power as such is rejected. As has been demonstrated, this attitude is based mainly on concerns about landscape values.

> III A positive attitude towards wind farms, which turns into a negative attitude as a result of the discussion surrounding the proposed construction of a wind farm.

> IV Resistance created by the fact that some construction plans are themselves faulty, without a rejection of the technology itself.

This classification shows that there is a wider range of possible motivations behind the opposition than a simple NIMBY attitude. But the reality seems to be even more complex. For example, the so-called not-in-any-back-yard case (II in the quotation above) could also have different features, with general support of wind farms but with a particular aversion towards wind farms close to urban settlements or, alternatively, towards wind farms situated in landscapes of particular beauty or natural interest.

A member of the opposition group at Westmill made clear his disapproval of some wind farms, and indicated his support for wind power only if the site is not close to a human settlement.

Excerpt 52

> I think that the varied impact of a wind farm on the environment and communities is such that we ought really at government level to have some clear guidelines that you "may not" have a wind farm within whatever distance is deemed reasonable of a big human community. To have one near a remote farmhouse where the farmer could be compensated for being moved away

somewhere else if his health is at risk, that is one thing, but to put it down near a community like this one is not, in my view, reasonable.

Another member of the opposition group made clear her opposition to turbines on the grounds of both spoiling the beauty of the countryside and negatively affecting the health of the local community. In this case, however, the first motive seems to be the primary one. The opposition group member expressed a not-in-any-backyard view, yet she is not against wind farms in general:

Excerpt 53

And there is even a place for wind turbines where they have got proper wind and where they can do a proper job, but there shouldn't be an imposition on them [local communities] ...

Another opponent, this time a councillor, showed concern about the impact of wind turbines on the health of the local community and about the impact on the landscape. He believed that wind power was not the best answer and he mentioned, to support his view, the intermittency of generation. Yet, despite being sceptical about wind power, at the end of the interview he expressed the following opinion:

Excerpt 54

R: If they were stocked offshore somewhere where just a passing ship sees them once in a while [laughs] and they [are] not a blight on the countryside then, you know, I think ... actually if they are shown to be efficient and working and a good investment I personally would go and invest in them. But ... it's NIMBYism, you know what NIMBYism is? Not in my back yard, you know? We don't want them here, we want them somewhere else [laughs], but the funny thing is that I actually approve of renewable energy sources, you know what I mean, I just don't ... I have this uneasy feeling that wind farms aren't the way, there must be other things.

I: *Mmm maybe this was not, for you, the appropriate place.*

R: I think that they could have been further away from the population, you know, there are fields out there, miles ... anywhere you can't see them ...

Another councillor who opposed the wind farm on the grounds of landscape damage said that he opposed it because at the time of the first application wind farms were rare and he believed that this was not an appropriate site. He also said that if wind farms were more prevalent, he would have accepted one in the locale, therefore suggesting a fairness issue—a sort of not-in-my-backyard unless it's in everyone's backyard.

Excerpt 55

I just didn't think that this was the right place at the right time. If we had most of England covered with wind farms, then fine, I don't have a problem having a wind farm here in Watchfield ...

He also thought that wind farms would be better suited to remote areas where local communities were experiencing unemployment. Although they would spoil the environment, their negative visual impacts would be overridden by the economic benefits of creating local jobs and possibly local revenue. Interestingly, the councillor also suggested that people living in remote and relatively poor rural communities did not fully appreciate the beauty of their environment and that, actually, they would appreciate local development and jobs more. This hypothesis is in accordance with post-materialist theories that suggest that diffused environmental concerns are only present in affluent societies (Inglehart, 1977; Diekmann and Franzen, 1999).

Excerpt 56

> It's probably better for some of these communities, some far-flung parts of this island, to actually have this kind of massive turbine and so on, because they don't realise … they don't recognise the environmental beauties that they have got, our land … And so, therefore, when they are [no longer] there but they've got jobs instead, it won't be such a worry to them, which is a very cynical thing to say, I know, but …

In conclusion, no pure NIMBY attitudes were found; in fact, three out of four of the interviewees expressed a "not in any back yard of my kind" sort of opposition. None of them rejected wind power in and of itself; rather, they suggested that wind turbines should be in places with specific characteristics, (e.g. far away from communities or offshore). The only exception was the last interviewee mentioned (Excerpt 58), who, arguably, came close to a NIMBY attitude, even though he was willing to accept wind farms if they were ubiquitous, and suggested that they should be located where local residents could appreciate the benefits they would provide. Therefore, each person interviewed expressed a sense of fairness and not the discriminatory selfishness generally attributed to the NIMBY syndrome. Westmill interviewees based their decision to oppose on issues that they mostly termed "spoiling the beauty of the landscape" or "threatening somehow the health of the local community", which they framed in arguments around fairness.

This finding is not unusual. Wolsink (2006, 2007), for example, holds a similar position. He claims (Wolsink, 2007, p. 1203): "the crucial fact in NIMBY issues is not egotism, nor any other personality trait, but fair decision making that does not cause any perceived injustice." Other authors (e.g. Gross, 2007; Botetzagias *et al.*, 2015; Bailey and Darkal, 2018) have similarly found instances where fairness has been central in motivating wind farm opposition.

Conclusions from the qualitative study

The research questions (Chapter 3) asked which factors influence the acceptability of wind farms and participation in wind farm co-operatives and how they relate to one another. These questions led to developing a literature review focusing on the costs and benefits that might persuade local residents to support a wind farm and become members of a wind farm co-operative. These costs/benefits were recognised as being of a different nature: economic, health and psychological, with their appreciation being influenced by personal resources (wealth and vicinity) and communication.

A strictly rational choice or, alternatively, a psychological framework, would possibly have lacked some interpretative elements useful to deepening the understanding of wind farm opposition. An integrated framework seems to be a useful interpretative tool in the light of literature that advocates such research development (Corraliza and Berenguer, 2000; Stern, 2000; Diekmann and Preisendörfer, 2003; de Groot and Steg, 2009; Perlaviciute and Steg, 2014; Steg *et al.*, 2014). Costs and benefits plausibly affect the first level of individuals' decision-making that might evidence a low-cost or low-benefit outcome. This leads to a second level of decisional process that includes other costs and benefits, which appear related to personal fulfilment—being consistent with the person's own attitudes chiefly, but not exclusively, towards the local and the global environment. Costs and benefits that were found to be relevant for the case of Westmill included: devaluation of property prices, perceived health risks (severe environmental stress due to noise and visual impact), free services for the local community, opportunity of benefiting from the co-operative's revenue, creation of local jobs, cognitive consistency with pro-environmental attitudes, aesthetic visual impact and increased levels of noise.

The analysis of the transcripts has suggested that personal resources such as income and proximity to the wind farm site might influence the importance, as perceived by residents, of costs and benefits of the proposed turbines. It was also observed that proximity and affluence might possibly increase appreciation of the costs in the pre-construction phase. Conversely, interviewees' greater distance from the site, and lower income, could possibly reduce their perception of the costs. Finally, communication was regarded by interviewees as a relevant issue: the inability to properly promote and communicate the aim of creating a community ownership scheme might have provoked a lack of credibility and an underestimation of the benefits of the project.

In the case of Watchfield, the local community's relative wealth, as mentioned by some interviewees, probably resulted in appreciation of the economic benefits that such a project could have provided locally being downplayed, while it increased appreciation of the risks of property devaluation. At the same time, the proximity of the local community to the wind farm site might have increased their perception of alleged health risks that the opposition attempted to magnify through its campaign. Furthermore, the opinion that the original proponent might have used the co-operative model to ease the planning process and build local consent could have played a negative role in the minds of some local residents.

One of the research questions asked whether the co-operative model of community-owned wind farms could overcome local opposition. The question was based on the hypothesis that a co-operative scheme could provide a financial benefit for the community, and could compensate for the disutility generated by its locally perceived impact (Toke, 2002). Interviewees indicated that the compensation was actually poor in respect of opponents, who would have excluded themselves from the share sale. Moreover, opponents would have felt betrayed and alienated by the part of the local community that supported the project and joined the co-operative. Some believed that the co-operative scheme had the potential to create an internal, possibly enduring, rift within the community, with the turbines serving as a constant reminder of a lost battle.

On a more positive note, there was widespread opinion that the co-operative scheme might, in differing conditions, be able to swing undecided individuals towards support, hence reducing the number of opponents.

Finally, the lack of examples of community-owned wind farm co-operatives was considered detrimental to the acceptance of this model at Westmill: people might have found it difficult to understand something that was, at the time, relatively unusual in the UK..

Another research question asked whether individuals perceived their citizen status as a source of moral obligation to protect the environment. The answers revealed differing levels of citizenship that previous scholarship has not considered. Hypothetically, following Dobson's (2003) categorisation, it was expected that a sense of moral obligation to act in support of the environment as a consequence of a sense of belonging to either the national community ("environmental citizenship") or the global community ("ecological citizenship") would be found. However, duty to one's country did not motivate a sense of responsibility towards the environment, while a sense of belonging to the local community emerged as a motive for action that was termed "local citizenship", i.e. the wish to protect the local community from the supposedly adverse impact of the wind farm on the local environment. Interviewees holding such positions also exhibited an emotional attachment to the local environment that induced a likely relationship with the concepts of "place attachment" and "place identity". Supporters, on the other hand, often referred to the importance of acting locally to tackle the global problem of climate change. Therefore, the Westmill case elicited a conflict between local and ecological citizenship consistent with what Warren *et al.* (2005) define as a conflict between local and global environmentalists in relation to onshore wind turbines.

Motivations to act on behalf of the local or the global community do not conflict with the subjective drives indicated earlier that form the basis of a rational choice attitudes integrated framework. Individuals are a part of their communities and what they recognise as positive for them, in relation to collective matters, is commonly viewed as positive for others in similar conditions. In doing so, individuals accrue a double advantage: first, they externally strengthen their argument by making it a collective issue and possibly gathering support; second, they are likely to nurture their self-esteem by promoting an image of themselves as champions of the public good, thereby, perhaps, satisfying an internal need (Maslow, 1987).

The interviews asked how stakeholders and citizens viewed social enterprises and co-operatives. Opinions were predominantly positive; nevertheless, three respondents (one activist opponent and two councillors who opposed the wind farm) exhibited scepticism about this kind of enterprise, which they viewed as less efficient than traditional organisations. Other interviewees expressed positive opinions towards co-operatives, and stressed the democratic and more socially just character of these enterprises. This issue was not perceived as central in the debate about the local wind farm; nevertheless, opponents suggested that the co-operative scheme was a sort of propagandistic move, or even a way to disguise the promoter's subjective interests.

Finally, the interview transcripts offered further evidence against the continued usage of the NIMBY label as argued by Wolsink (2007). In fact, the egoistic motives assumed by the term NIMBY contrast with the sense of fairness displayed in wind farm opponents' answers: they rejected the possibility of installing the turbines elsewhere. This typology of opponents, which is not captured in Wolsink's (2007) classification, represented the vast majority of opponents' answers that I collected. Hence, despite its consistency with previous research (Wolsink, 2007), this analysis supports a further type of opposition, whose motivation appears to be a sense of fairness that emerges not only as a desire not to be discriminated against, but also to reject "damaging" wind farms elsewhere that might be equally close to local communities.

In conclusion, the analysis supported the framework proposed in Chapter 2, which integrates rational choice and socio-psychological theories. This appears to be a potentially useful theoretical approach for deepening our understanding of opposition to wind farms; it probes the interaction between attitudes, perceived costs and benefits and their relation to individuals' personal resources, while confirming the relevance of several issues identified in previous research on wind farms.

Notes

1 Midcounties Co-operative is the UK's largest independent co-operative, with over 700,000 members. It operates a range of businesses in food, travel, healthcare, funerals, childcare, energy, post offices, flexible benefits and telecoms. It generated a revenue of £1.18 billion in the financial year 2018–2019. See www.midcounties.coop/siteassets/reports/18_19_Annual_Report.
2 The Westmill Wind Farm weblog has since been removed from the internet, hence it is not possible to reference it. The blog was copied in 2006 and used as material for the qualitative analysis, along with the interview transcripts.

References

Bailey, I. and Darkal, H. 2018. (Not) talking about justice: Justice self-recognition and the integration of energy and environmental–social justice into renewable energy siting. *Local Environment*, 23, 335–351.
Bonaiuto, M., Carrus, G., Martorella, H., Bonnes, M. and Mirilia, B. 2002. Local identity processes and environmental attitudes in land use changes: The case of natural protected areas. *Journal of Economic Psychology*, 23, 631–653.

Botetzagias, I., Malesios, C., Kolokotroni, A. and Moysiadis, Y. 2015. The role of NIMBY in opposing the siting of wind farms: Evidence from Greece. *Journal of Environmental Planning and Management*, 58, 229–251.

Corraliza, J. A. and Berenguer, J. 2000. Environmental values, beliefs, and actions: A situational approach. *Environment and Behavior*, 32, 832–848.

de Groot, J. I. M. and Steg, L. 2009. Mean or green: Which values can promote stable pro-environmental behavior? *Conservation Letters*, 2, 61–66.

Devine-Wright, P. 2005. Beyond NIMBYism: Towards an integrated framework for understanding public perceptions of wind energy. *Wind Energy*, 8, 125–139.

Devine-Wright, P. 2009. Rethinking NIMBYism: The role of place attachment and place identity in explaining place-protective action. *Journal of Community & Applied Social Psychology*, 19, 426–441.

Devine-Wright, P. and Howes, Y. 2010. Disruption to place attachment and the protection of restorative environments: A wind energy case study. *Journal of Environmental Psychology*, 30, 271–280.

Devine-Wright, P. and Wiersma, B. 2020. Understanding community acceptance of a potential offshore wind energy project in different locations: An island-based analysis of "place-technology fit". *Energy Policy*, 137, 111086. Available at https://doi.org/10.1016/j.enpol.2019.111086.

Diekmann, A. and Franzen, A. 1999. The wealth of nations and environmental concern. *Environment and behavior*, 31, 540–549.

Diekmann, A. and Preisendörfer, P. 2003. Green and greenback: The behavioral effects of environmental attitudes in low-cost and high-cost situations. *Rationality and Society*, 15, 441–472.

Dobson, A. 2003. *Citizenship and the environment*. Oxford: Oxford University Press.

DPDS Consulting Group. 2002. *Environmental impact assessment: Non-technical summary*. Available at: www.tvenergy.org/pdfs/Main%20Text.pdf.

Gardner, G. T. and Stern, P. C. 2002. *Environmental problems and human behaviour*. Boston, MA: Pearson Custom Publishing.

Gross, C. 2007. Community perspectives of wind energy in Australia: The application of a justice and community fairness framework to increase social acceptance. *Energy Policy*, 35, 2727–2736.

Halliday, J. 1993. Wind energy: An option for the UK? *IEE Proceedings A (Science, Measurement and Technology)*, 140(1), 53–62.

Inglehart, R. 1977. *The silent revolution: Changing values and political styles among Western publics*. Princeton, NJ: Princeton University Press.

Kyle, G. T., Mowen, A. J. and Tarrant, M. 2004. Linking place preferences with place meaning: An examination of the relationship between place motivation and place attachment. *Journal of Environmental Psychology*, 24, 439–454.

Lothian, A. 2008. Scenic perceptions of the visual effects of wind farms on south Australian landscapes. *Geographical Research*, 46, 196–207.

Maslow, A. H. 1987. *Motivation and personality*. New York: Harper & Row.

McLean, I. and McMillan, A. 2009. *The concise Oxford dictionary of politics*. Oxford: Oxford University Press.

Molnarova, K., Sklenicka, P., Stiborek, J., Svobodova, K., Salek, M. and Brabec, E. 2012. Visual preferences for wind turbines: Location, numbers and respondent characteristics. *Applied Energy*, 92, 269–278.

ONS. 2001. *Census 2001*. Office for National Statistics, London. Available at: www.statistics.gov.uk/census.

Perlaviciute, G. and Steg, L. 2014. Contextual and psychological factors shaping evaluations and acceptability of energy alternatives: Integrated review and research agenda. *Renewable and Sustainable Energy Reviews*, 35, 361–381.

Reeve, A. 2009. Citizenship. In McLean. I. and McMillan, A. (eds.) *The concise Oxford dictionary of politics* (3rd edn.). Oxford: Oxford University Press.

Rippetoe, P. A. and Rogers, R. W. 1987. Effects of components of protection-motivation theory on adaptive and maladaptive coping with a health threat. *Journal of Personality and Social Psychology*, 52(3), 596–604.

Steg, L., Bolderdijk, J. W., Keizer, K. and Perlaviciute, G. 2014. An integrated framework for encouraging pro-environmental behaviour: The role of values, situational factors and goals. *Journal of Environmental Psychology*, 38, 104–115.

Stern, P. C. 2000. New environmental theories: Toward a coherent theory of environmentally significant behavior. *Journal of Social Issues*, 56, 407–424.

Stern, P. C., Dietz, T., Abel, T. D., Guagnano, G. A. and Kalof, L. 1999. A value-belief-norm theory of support for social movements: The case of environmentalism. *Human Ecology Review*, 6, 81–97.

Sterzinger, G., Beck, F. and Kostiuk, D. 2003. *The effect of wind development on local property values*. Washington: Renewable Energy Policy Project.

Toke, D. 2002. Wind power in UK and Denmark: Can rational choice help explain different outcomes? *Environmental Politics*, 11, 83–100.

TV Energy. 2004. *TV Energy case study: Westmill Wind Farm, Oxfordshire*. Available at: www.tvenergy.org/pdfs/westmil-web-case-study.pdf.

van Veelen, B. and Haggett, C. 2016. Uncommon ground: The role of different place attachments in explaining community renewable energy projects. *Sociologia Ruralis*, 57, 533–554.

Verba, S., Schlozman, K. L. and Brady, H. E. 1995. *Voice and equality: Civic voluntarism in American politics*. Cambridge, MA: Harvard University Press.

Warren, C., Lumsden, C., O'Dowd, S. and Birnie, R. 2005. "Green on green": Public perceptions of wind power in Scotland and Ireland. *Journal of Environmental Planning and Management*, 48, 853–875.

Weston, J. 2002. *Public opinion on the impact of the proposed Westmill Farm wind farm Watchfield Oxfordshire*. Impact Assessment Unit, School of Planning, Oxford Brookes University, UK.

Wolsink, M. 2000. Wind power and the NIMBY-myth: Institutional capacity and the limited significance of public support. *Renewable Energy*, 21, 49–64.

Wolsink, M. 2006. Invalid theory impedes our understanding: A critique on the persistence of the language of NIMBY. *Transactions of the Institute of British Geographers*, 31, 85–91.

Wolsink, M. 2007. Wind power implementation: The nature of public attitudes: Equity and fairness instead of "backyard motives". *Renewable and Sustainable Energy Reviews*, 11, 1188–1207.

Xu, Z. and Zhang, J. 2016. Antecedents and consequences of place attachment: A comparison of Chinese and Western urban tourists in Hangzhou, China. *Journal of Destination Marketing & Management*, 5, 86–96.

5   A survey of four Scottish proposed wind farms

This chapter presents the findings of a postal survey of residents living within a 10 km radius of four proposed wind farm sites in Scotland—Nigg Hill, Cushnie, Meikle Carewe and Bracco—whose cases were presented in Chapter 3. It first details the demographic composition of the sample before exploring the relationships between demographic characteristics and opinions about the proposed wind farms. The chapter then discusses the analysis of the survey and, in particular, bivariate and multivariate correlations, offering a quantitative understanding of how the factors which influence acceptance of wind farms and participation in a co-operative scheme are related to one another.

Personal resources and demographic variables

Response rate per site

A total of 315 questionnaires were returned from the four sites; however, due to their random and inconsistent completion, two questionnaires were excluded from the analysis.

The largest number of questionnaires were returned from Cushnie, closely followed by Nigg Hill and Meikle Carewe, while Bracco had a substantially lower rate of return. The rates of return were 45 per cent for Cushnie, 32 per cent for Nigg Hill, 30 per cent for Meikle Carewe and 17 per cent for Bracco (see Table 5.1).

Bracco's low rate of return might be indicative of the area's socio-environmental characteristics. Bracco is the most deprived of the areas surveyed and has an average SIMD (Scottish Index of Multiple Deprivation) ranking of 2162 versus 2892, 4972 and 5414 for Nigg Hill, Cushnie and Meikle Carewe, respectively. Social deprivation has been highlighted as the chief element negatively affecting political engagement (The Electoral Commission, 2005). van der Horst and Toke (2010) found that deprivation was related to increased planning approval of wind farm proposals, thereby suggesting that in socially deprived contexts, people have fewer resources to lobby against wind farm developments.

Nevertheless, Nigg Hill, which is also a relatively deprived area, returned nearly double the number of questionnaires as Bracco. Two potential factors

Table 5.1 Distribution of respondents across surveyed sites

		Frequency	*Percent*	*Valid percent*
Valid	Cushnie	113	35.9	36.1
	Nigg Hill	82	26.0	26.2
	Meikle Carewe	75	23.8	24.0
	Bracco	43	13.7	13.7
	Total	*313*	*99.4*	*100.0*
Missing	System	2	0.6	–
Total		*315*	*100.0*	–

may explain Nigg Hill's higher participation rate. First, it is possible that as the Nigg Hill area is more rural than Meikle Carewe (the urban rural classification average is 2.91 for Meikle Carewe compared to 5.16 for Nigg Hill), the community might believe that the proposed wind farm's potential impact on the integrity of their environment would be greater. Second, Nigg Hill's population is older, which in itself facilitates greater political engagement (The Electoral Commission, 2005). Available data from the 2001 Census show that roughly 18.56 per cent of the area's population are pensioners, compared to 14.45 per cent in the Meikle Carewe area and 15.75 per cent in the Bracco area (ONS 2001).

The majority of respondents across all areas (56.2 per cent) lived within a 0–5 km radius of the proposed wind farm, while 43.5 per cent lived within 5–10 km of the site. The difference is not drastic, but it appears reasonable if we consider that residents who lived closer to the proposed wind farm might feel more affected by its impacts. This finding is consistent with other surveys, in which it has been suggested that physical proximity to the site might have created a self-selection bias that led to higher numbers of opponents returning questionnaires (Jones and Eiser, 2009), or that distance from residential areas was a decisive factor influencing respondents (a matter discussed in Chapter 2).

Interestingly, more questionnaires were returned from the proposed co-operative wind farm schemes, Nigg Hill and Cushnie (62.3 per cent), than from commercial scheme sites (37.7 per cent). The reasons for different rates of return appear to hold for explaining this difference as well, although one additional explanation could be that co-operative schemes forced a wider debate about the wind farm project that piqued public interest and increased the likelihood that residents would develop an opinion about the matter.

Education

The majority of respondents stated that "secondary school" (36 per cent) or "professional/vocational" education (30 per cent) was their highest level of education. "Graduates" and "postgraduates" comprised 17.5 per cent and 14.5 per cent of respondents, respectively (see Figure 5.1). These educational levels

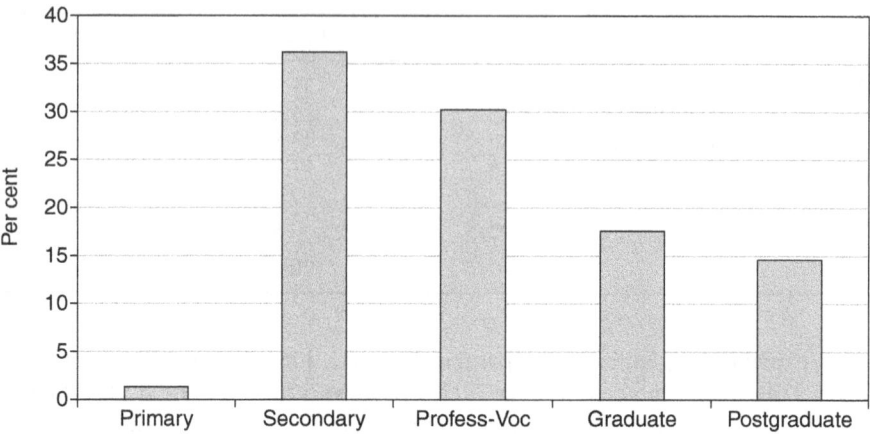

Figure 5.1 Level of education declared.

may be representative of the population as a whole; unfortunately, it was not possible to locate any data that could confirm this.

The level of education varies from one wind farm site to another (Table 5.2). It is striking that 75 per cent of all respondents who had attained a primary school level of education only were from the Bracco site, while 45 per cent of those who said they had attained a university degree, and 40 per cent who stated they had a postgraduate degree, were located in Cushnie. This is consistent with the levels of SIMD presented in Chapter 3 for the areas surveyed, which showed higher levels of multiple deprivation for the Bracco and Nigg Hill sites.

Income

Nearly one-quarter (26 per cent) of respondents declared a household gross income bracket of £30,000–49,000, while the majority (55.3 per cent) have a household income of £30,000 or more (Table 5.3).

Once again, the SIMD score of the single locations were reflected in the household incomes respondents declared; however, response rates varied widely, from just 34 respondents in the Bracco area to 104 from the Cushnie area. In the Bracco area, nearly one-third of respondents declared a household income of less than £10,000 per year, while almost one-quarter of respondents from the Meikle Carewe area declared a household income of more than £80,000 per year (Table 5.4).

Twelve per cent of respondents declined to provide information about their income, presumably because they considered it private. This omission was compounded by a lack of information on the number of individuals residing in a household, making it possible to estimate the gross income per person for just 236 households. The single largest gross income per person value is £20,000,

Table 5.2 Level of education declared and wind farm sites

		Level of education					
		Primary school	Secondary School	Professional/ vocational	Graduate	Postgraduate	Total
Name of wind farm							
Cushnie	Count	1	37	29	24	18	109
	% within [name of wind farm]	0.9%	33.9%	26.6%	22.0%	16.5%	100.0%
	% within level of education	25.0%	33.9%	31.9%	45.3%	40.9%	36.2%
Nigg Hill	Count	0	33	23	9	12	77
	% within [name of wind farm]	0.0%	42.9%	29.9%	11.7%	15.6%	100.0%
	% within level of education	0.0%	30.3%	25.3%	17.0%	27.3%	25.6%
Meikle Carewe	Count	0	19	30	13	12	74
	% within [name of wind farm]	0.0%	25.7%	40.5%	17.6%	16.2%	100.0%
	% within level of education	0.0%	17.4%	33.0%	24.5%	27.3%	24.6%
Bracco	Count	3	20	9	7	2	41
	% within [name of wind farm]	7.3%	48.8%	22.0%	17.1%	4.9%	100.0%
	% within level of education	75.0%	18.3%	9.9%	13.2%	4.5%	13.6%
Total	Count	4	109	91	53	44	301
	% within [name of wind farm]	1.3%	36.2%	30.2%	17.6%	14.6%	100.0%
	% within level of education	100.0%	100.0%	100.0%	100.0%	100.0%	100.0%

Table 5.3 Gross household income declared in British pounds (GBP)

		Frequency	Percent	Valid percent	Cumulative percent
Income	Under 10,000	41	13.0	14.9	14.9
	10,000–19,999	39	12.4	14.2	29.1
	20,000–29,999	43	13.7	15.6	44.7
	30,000–49,999	72	22.9	26.2	70.9
	50,000–79,999	47	14.9	17.1	88.0
	80,000 plus	33	10.5	12.0	100.0
	Total	*275*	*87.4*	*100.0*	–
Missing	System	38	12.1	–	–
		2	0.6	–	–
	Total	*40*	*12.7*	–	–
Total		*315*	*100.0*	–	–

Table 5.4 Declared household income by wind farm site

Name of wind farm		Household income						
		Under 10,000	10,000–19,999	20,000–29,999	30,000–49,999	50,000–79,999	80,000 plus	Total
Cushnie	Count	13	16	17	29	17	12	104
	% by wind farm	12.5%	15.4%	16.3%	27.9%	16.3%	11.5%	100.0%
	% within household income	31.7%	41.0%	39.5%	40.3%	36.2%	36.4%	37.8%
Nigg Hill	Count	14	8	11	22	12	5	72
	% by wind farm	19.4%	11.1%	15.3%	30.6%	16.7%	6.9%	100.0%
	% within household income	34.1%	20.5%	25.6%	30.6%	25.5%	15.2%	26.2%
Meikle Carewe	Count	4	10	9	14	12	16	65
	% by wind farm	6.2%	15.4%	13.8%	21.5%	18.5%	24.6%	100.0%
	% within household income	9.8%	25.6%	20.9%	19.4%	25.5%	48.5%	23.6%
Bracco	Count	10	5	6	7	6	0	34
	% by wind farm	29.4%	14.7%	17.6%	20.6%	17.6%	0.0%	100.0%
	% within household income	24.4%	12.8%	14.0%	9.7%	12.8%	0.0%	12.4%
Total	Count	41	39	43	72	47	33	275
	% by wind farm	14.9%	14.2%	15.6%	26.2%	17.1%	12.0%	100.0%
	% within household income	100.0%	100.0%	100.0%	100.0%	100.0%	100.0%	100.0%

which includes 17.8 per cent of the respondents who declared both an income bracket and the number of household members. Most respondents (49 per cent) reported that they resided in a two-person household.

Knowledge about wind energy

A range of questions assessed respondents' knowledge of the characteristics of wind power, namely, pollution, the cost of producing electricity through wind generation, intermittency and whether wind was considered a renewable source of energy. The overwhelming majority of respondents (89.5 per cent) said that wind generation is less polluting than coal-fired power stations. Fifty-one per considered wind power a cheaper means of producing electricity than coal (Figure 5.2), while 60 per cent believed that wind did not produce a steady stream of electricity regardless of the location. Respondents from the Bracco area showed a different pattern of answers compared with the other sites: over 40 per cent answered, "I don't know" in response to the question regarding whether wind turbines produce a steady stream of electricity (Figure 5.3).

The vast majority of all respondents (88 per cent) considered wind to be a type of renewable energy, but about 20 per cent of respondents from the Bracco area answered, "I don't know".

Community scheme awareness

Respondents were asked whether they were aware of a community scheme being proposed for the wind farm; the vast majority (87 per cent) answered that

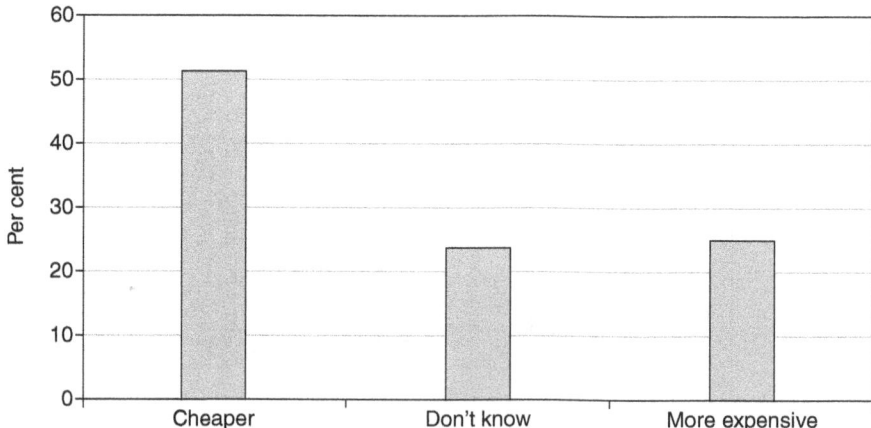

Figure 5.2 Responses to the question: "Is electricity produced by wind turbines cheaper or more expensive to produce than electricity produced by other means such as coal-fired power stations?".

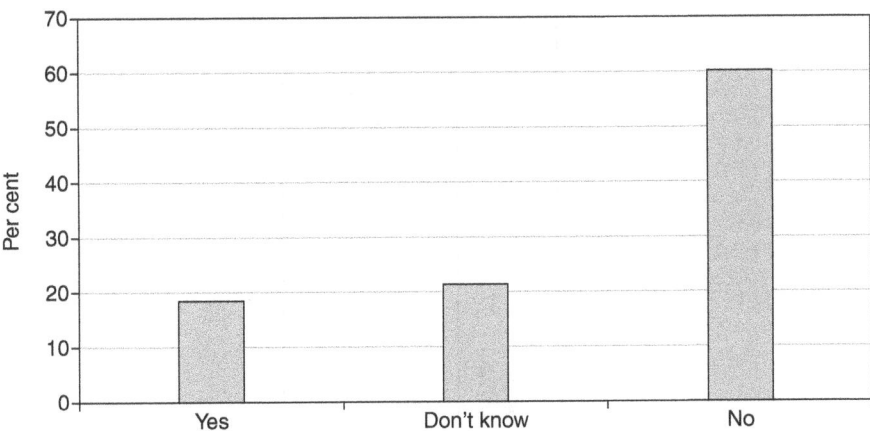

Figure 5.3 Responses to the question: "Whatever the location, do wind turbines produce a steady stream of electricity?".

they were not aware (Table 5.5). The number of "not aware" respondents was highest in the Meikle Carewe area (one of the two purely commercial wind farms), where 97 per cent responded in this way. On the contrary, the lowest number of "not aware" respondents (75 per cent) were from the Nigg Hill area (one of the two co-operative schemes). In the case of Nigg Hill, 14 per cent reported awareness of a scheme that included the co-operative, while 10 per cent of Cushnie respondents (the other co-operative scheme case surveyed) reported knowing about a co-operative scheme.

Awareness and opinion about the locally proposed wind farm

Respondents were asked about their awareness of the proposed wind farm. Percentages varied across the proposed sites, with Cushnie and Nigg Hill respondents claiming to be very aware, while Meikle Carewe respondents were almost equally divided between being aware and not aware. In Bracco, the vast majority of respondents were not aware. Again, a possible explanation could be that in areas where a co-operative scheme was proposed, the debate over their implementation raised public awareness (Figure 5.4).

When respondents were asked whether they could see the wind farm site from their house, a relative majority of respondents (46 per cent) said "No", while 23 per cent said they did not know the site's location (Figure 5.5).

Responses to the question about whether the site could be seen varied between the areas surveyed, with nearly 60 per cent of respondents from the Nigg Hill area answering "Yes" and over 60 per cent from the Bracco area responding, "I don't know where the site is" (Figure 5.6).

Table 5.5 Awareness of the proposed community scheme for the local wind farm

		Community fund (%)	Community wind farm co-op (%)	Different scheme (%)	Not aware of any (%)	Fund + co-op (%)	Fund + co-op + community turbine (%)	Fund + other (%)	Total (%)
Name of wind farm	Cushnie	1.8	7.3	0.9	87.3	2.7	–	–	100.0
	Nigg Hill	11.2	3.8	–	75.0	6.2	3.8	–	100.0
	Meikle Carewe	1.4	–	–	97.3	1.4	–	–	100.0
	Bracco	–	2.4	2.4	92.7	–	–	2.4	100.0
	Total	*3.9*	*3.9*	*0.7*	*87.2*	*3.0*	*1.0*	*0.3*	*100.0*

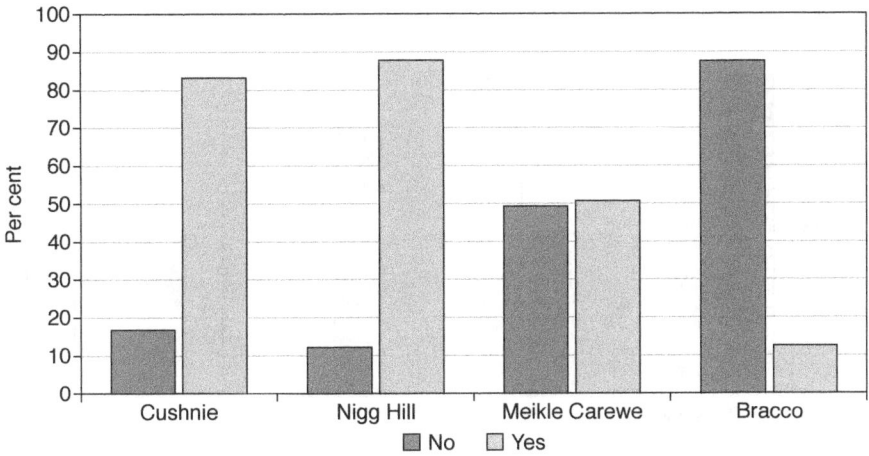

Figure 5.4 Responses to the question: "Are you aware that the [...] wind farm has been proposed in your local area?".

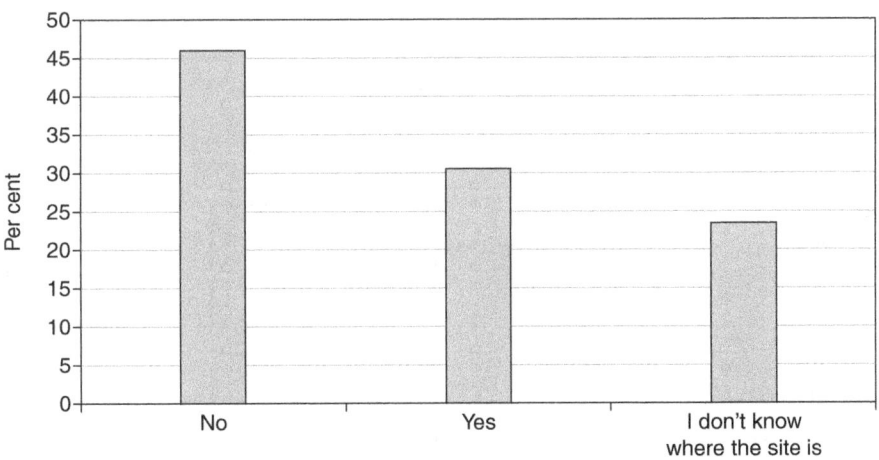

Figure 5.5 Responses to the question: "Can you see from your home [...], the site of your proposed local wind farm?".

Of those that answered "No" to the question whether they could see the wind farm site from their house, a relative majority (36.5 per cent) reported that they saw the site "sometimes", while 24.5 per cent said they saw it "often" (Figure 5.7).

Opinion about the locally proposed wind farm

One of the study's important variables concerned respondents' opinions regarding the locally proposed wind farm. Across all respondents, 28.2 per cent

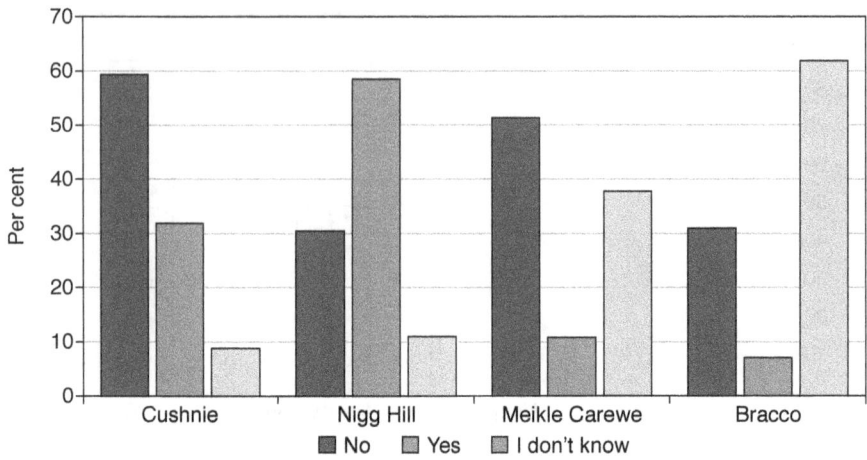

Figure 5.6 Responses to the question: "Can you see from your home [...], the site of your proposed local wind farm?"—by area surveyed.

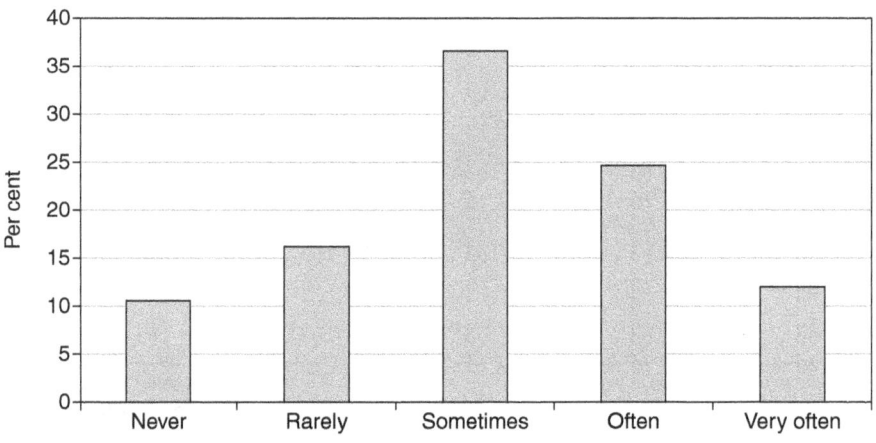

Figure 5.7 Frequency of seeing the wind farm site among respondents who cannot see the wind farm from home.

said "agree" followed by 26.9 per cent who replied "neither agree nor disagree", 20.1 per cent who responded "strongly disagree", followed by 14.6 per cent for "strongly agree" and 10.1 per cent for "disagree" (Figure 5.8).

Opinions about proposed wind farms vary widely across sites. A majority of respondents near Cushnie and Meikle Carewe "agree", while near Bracco most replied "neither agree nor disagree" and in Nigg Hill most "strongly disagree".

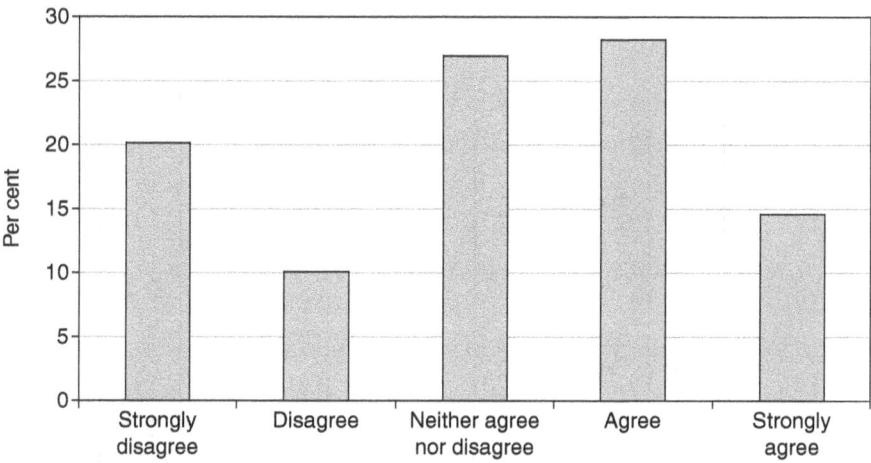

Figure 5.8 Responses to the question: "What do you think of the presence of this wind farm in your area?".

Table 5.6 provides a breakdown of the frequencies. With the exception of Bracco, the total number of respondents in agreement (the sum of "strongly agree" and "agree" answers) is the largest number in each site when compared with those undecided and those who disagree.

The questionnaire provided a blank space for respondents to specify their motives for the answer they provided to the above question. These are presented in Tables 5.7–5.9, which are divided into three groups of respondents: group 1 includes those who "strongly disagree" and who "disagree"; group 2, includes those who "neither agree nor disagree"; and group 3 includes those who "strongly agree" and who "agree". The results were obtained by grouping all indicated motives that did not achieve 9 per cent of frequency in at least one group of respondents (supporters, undecided and opposers) within the "other" category.

As Tables 5.7, 5.8, and 5.9 show, supporters (Table 5.7) mainly cited the fact that wind is a renewable energy source (18.2 per cent), which they consider "better than other sources" of energy (11.4 per cent), as the reason for their support. Those undecided (Table 5.8), mostly define themselves as such because of a "lack of sufficient information" (26.5 per cent). Opponents (Table 5.9) cited "visual impact" (41 per cent), followed by "inappropriate location" (18 per cent) as their main motives.

Attitudinal factors

A range of attitudinal questions evaluated perceived costs and benefits, environmental citizenship, environmental attitudes and place attachment.

Table 5.6 Opinion regarding the locally proposed wind farm across surveyed sites

Name of wind farm	Strongly disagree (%)	Disagree (%)	Neither agree nor disagree (%)	Agree (%)	Strongly agree (%)	Total (%)
Cushnie	20.5	14.3	19.6	29.5	16.1	100.0
Nigg Hill	28.0	6.1	23.2	25.6	17.1	100.0
Meikle Carewe	17.8	9.6	31.5	32.9	8.2	100.0
Bracco	7.3	7.3	46.3	22.0	17.1	100.0
Total	*20.1*	*10.1*	*26.9*	*28.2*	*14.6*	*100.0*

Table 5.7 Motives for opinions of supporters of the locally proposed wind farm

	Motive for opinion						Total
	Benefits the environment	Renewable/ sustainable energy	Not bothered	Other	Better than other sources	Energy security	
Supporters Count	13	24	1	69	15	10	132
%	9.8	18.2	0.8	52.3	11.4	7.6	100.0

Table 5.8 Motives for opinions of undecided about the locally proposed wind farm

	Motive for opinion								Total
	Negative visual impact	Benefits the environment	Renewable/ sustainable energy	Inappropriate location	Not bothered	Other	Better than other sources	Lacking sufficient information	
Undecided Count	1	2	4	1	5	47	1	22	83
%	1.2	2.4	4.8	1.2	6.0	56.6	1.2	26.5	100.0

Table 5.9 Motives for opinions of opponents of the locally proposed wind farm

	Motive for opinion					Total
	Negative visual impact	Inefficient/ intermittency	Inappropriate location	Other	Lacking sufficient information	
Opposers Count	38	10	17	26	2	93
%	40.9	10.8	18.3	28.0	2.2	100.0

Environmental citizenship

Questions concerning environmental citizenship (items 18–18.3 in the question-naire, see Appendix A) aimed to ascertain whether respondents felt a responsibility to act pro-environmentally and, if they did, at which level. Respondents were roughly divided between the statements, "We all have to do something to protect the global environment because we all share planet earth", with which 46.5 per cent of respondents agreed, and "We all have to do something to protect our local environment, because it's the place where we live with our families, in our communities", with which 38 of respondents agreed (Figure 5.9). These responses suggest a greater attachment to local and global environments and a disregard of the national dimension, or the option of rejecting any responsibility for environmental matters which, in the fourth item (18.3—"it's a matter for the government, not us, to take care of the environment"), was considered to be the sole responsibility of the government.

Pro-environmental attitudes vs pro-economy attitudes

This question asked respondents to give a preference to one of four statements representing a hypothetical trade-off between economic growth and environmental protection (items 19–19.3 of the questionnaire, see Appendix A). The majority of respondents (54.5 per cent) chose the statement, "The economy should be the priority of the government but this should not damage the environment", while 34.5 per cent selected the statement, "The environment should be the priority of the government but this should not damage the economy" (Figure 5.10).

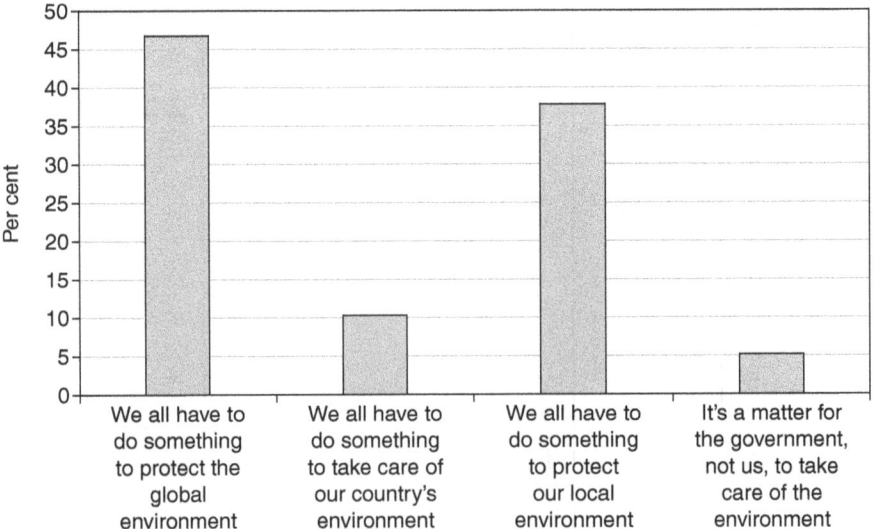

Figure 5.9 Perceived environmental responsibility and level of action.

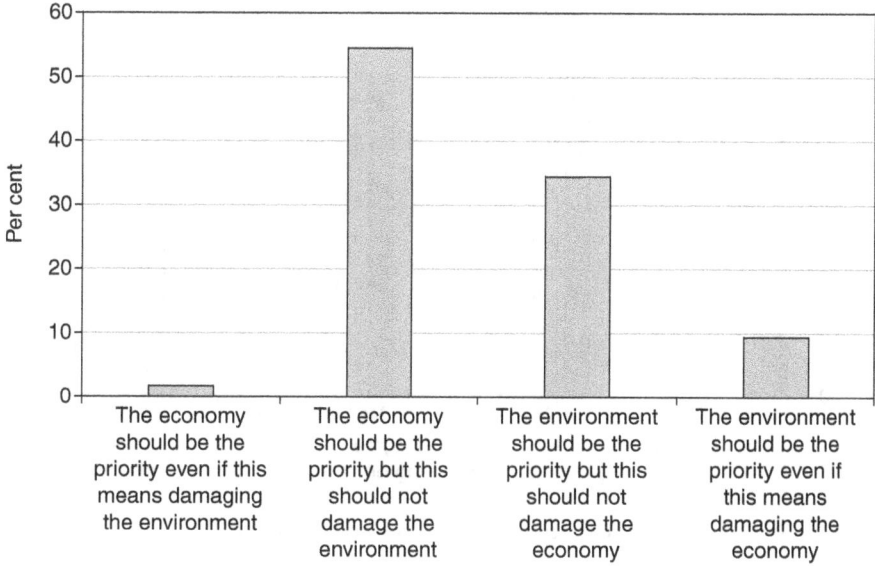

Figure 5.10 Preferences between prioritising economic growth and environmental protection.

Place attachment

Respondents were presented with two statements asking about place attachment. The first, "I like how my area looks", aimed to capture attachment to the physical environment (item 20 in the questionnaire, see Appendix A). The results are presented here with a bar chart detailing values across wind farm locations (Figure 5.11). Across all locations, with the exception of Bracco, "strongly agree" was by far the preferred category. In Bracco, most respondents stated that they "agree". Bracco also exhibited much larger proportions of "neither agree nor disagree" and "disagree" when compared with the other locations, where such replies were given in negligible numbers.

When asked about attachment to the social environment, with the statement "I like my community" (item 20.1 of the questionnaire, see Appendix A), the majority of respondents answered that they "strongly agree". Again, the pattern of responses for Bracco residents differed, with a prevalence of "agree" and comparatively higher proportions of "neither agree nor disagree" and "disagree" (Figure 5.12).

Perceived local costs and benefits

Respondents were asked whether there would be local benefits or disadvantages from the proposed wind farm and predominantly replied "more benefits than disadvantages" (42 per cent) (Figure 5.13). About 34 per cent believed that the

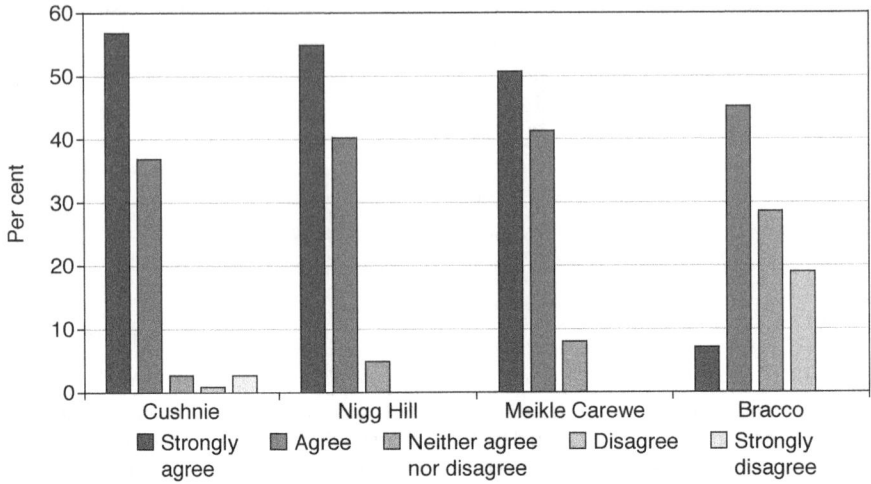

Figure 5.11 Responses to the statement: "I like how my area looks".

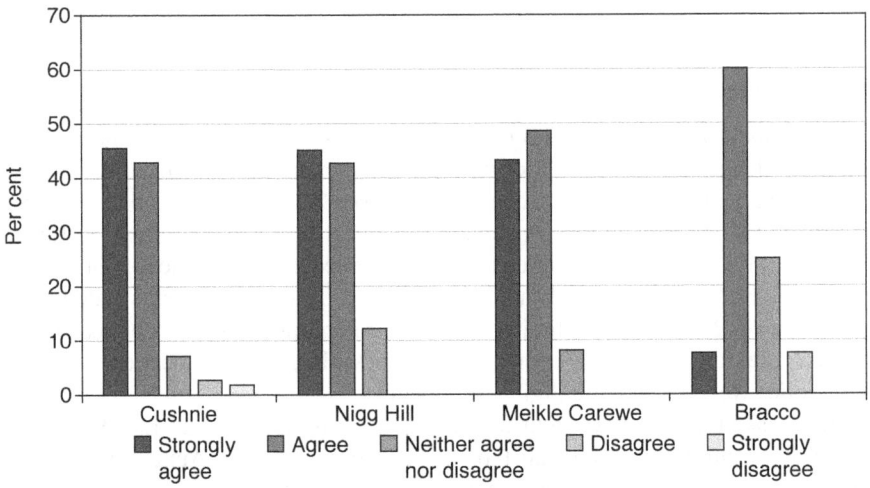

Figure 5.12 Responses to the statement: "I like my community".

wind farm would have brought "more disadvantages than benefits" or "many disadvantages and no benefits", compared with 46 per cent who thought the opposite. Looking at individual results, it is striking that slightly more respondents in the area of Nigg Hill believed that the wind farm would have more disadvantages than benefits, although the difference in percentage terms with those thinking the opposite is not substantial, 43 per cent compared to 41 per cent.

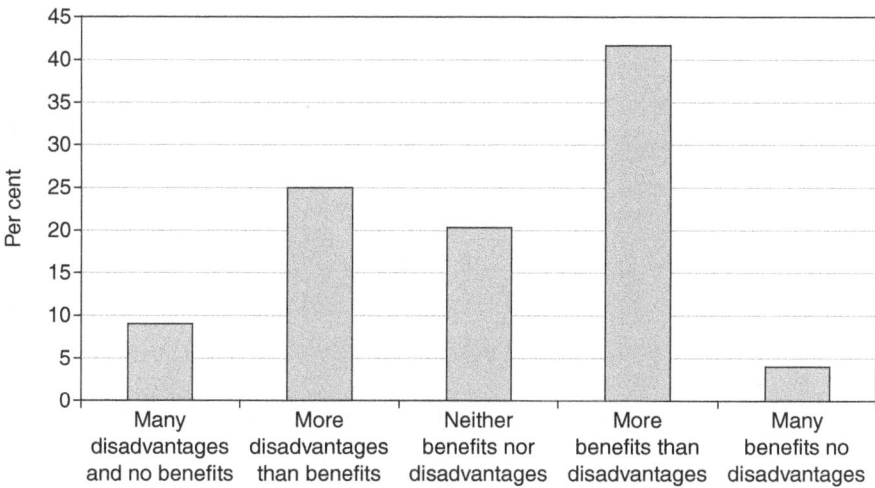

Figure 5.13 Responses to the statement asking whether the wind farm will bring local benefits or disadvantages.

Items 11–11.8 of the questionnaire asked specifically about nine benefits and costs associated with the wind farm in the event that it was built (see Appendix A).

Health impact

The first of these items regarded negative health consequences associated with the presence of the wind farm. Forty-one per cent of respondents disagreed with the statement, "The wind farm will harm the health of my community" (Figure 5.14). Combined with respondents who strongly disagreed, the majority (66 per cent) of respondents did not believe that the proposed wind farm might harm the health of the local community. Across all areas most respondents disagreed with the statement, with the exception of the Bracco area where nearly 40 per cent of respondents replied, "neither agree nor disagree".

Climate change impact

Respondents were asked about the contribution of the wind farm to climate change; 56 per cent of respondents "agreed" or "strongly agreed" with the statement, "The wind farm will help to fight climate change" (Figure 5.15).

Visual impact

Asked about the negative visual impact of the wind farm, 26.5 per cent of respondents neither agreed nor disagreed with the statement, "The wind farm

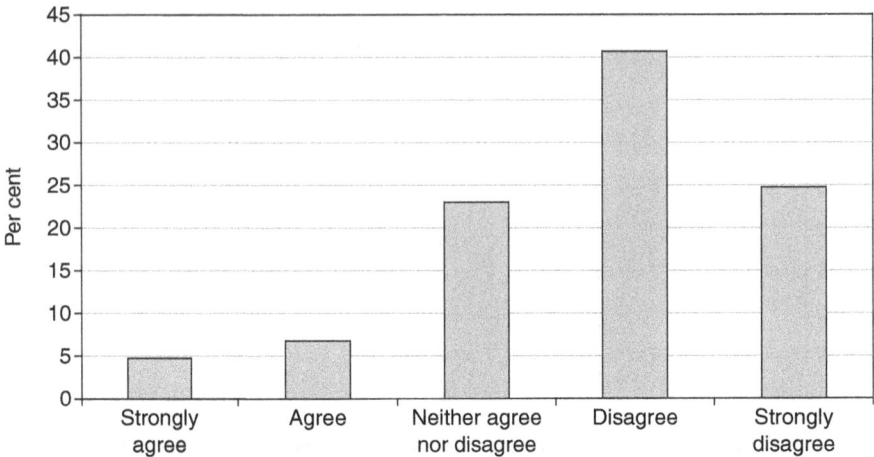

Figure 5.14 Responses to the statement: "The wind farm will harm the health of my community".

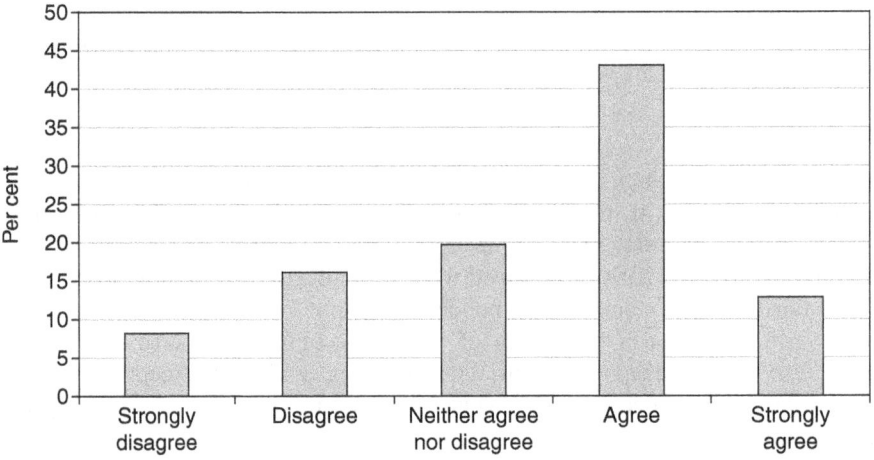

Figure 5.15 Responses to the statement: "The wind farm will help to fight climate change".

will look bad on the landscape". Nevertheless, an equally large number of respondents (26 per cent) strongly agreed with the statement; when their numbers are added to those who agree, the combined percentage is 41.5 per cent. This, therefore, indicates that most respondents felt that the wind farm would have a negative visual impact on their landscape. In the area of Bracco, those replying "neither agree nor disagree" were notably higher than in the other areas and the percentage reached almost 40 per cent (Figure 5.16).

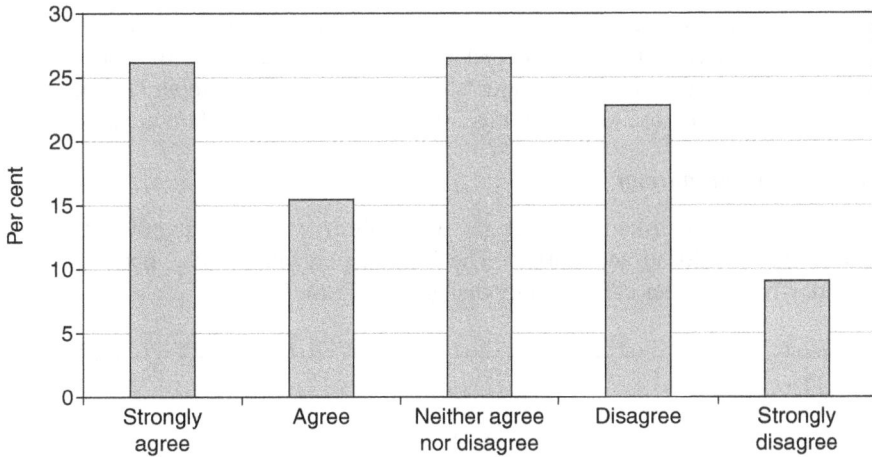

Figure 5.16 Responses to the statement: "The wind farm will look bad on the landscape".

Impact on the local economy

Respondents were asked whether the wind farm would improve the local economy. They largely replied "neither agree nor disagree" (35 per cent); nevertheless, 37 per cent of respondents indicated either "strongly disagree" or "disagree" (Figure 5.17). The picture changes across sites. The highest percentage of individuals (38 per cent) that believed the proposed wind farm would improve the local economy was registered in Nigg Hill, one of the two deprived areas surveyed.

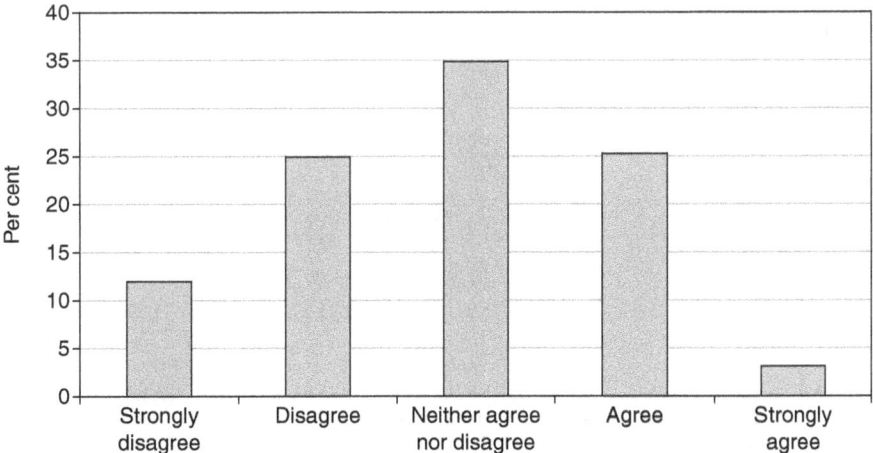

Figure 5.17 Responses to the statement: "The wind farm will improve the local economy".

Impact on property prices

When asked about the wind farm's negative impact on local property values, respondents largely replied "neither agree nor disagree" (36.5 per cent); however, the combined "agree" and "strongly agree" replies total 37.5 per cent (Figure 5.18). This rose to a little over 50 per cent in the Nigg Hill area.

Impact on local tourism

Respondents were asked whether the wind farm was a local environmental feature that would attract tourists. The highest percentage, 38.5 per cent, disagreed, while 28.5 per cent strongly disagreed (Figure 5.19).

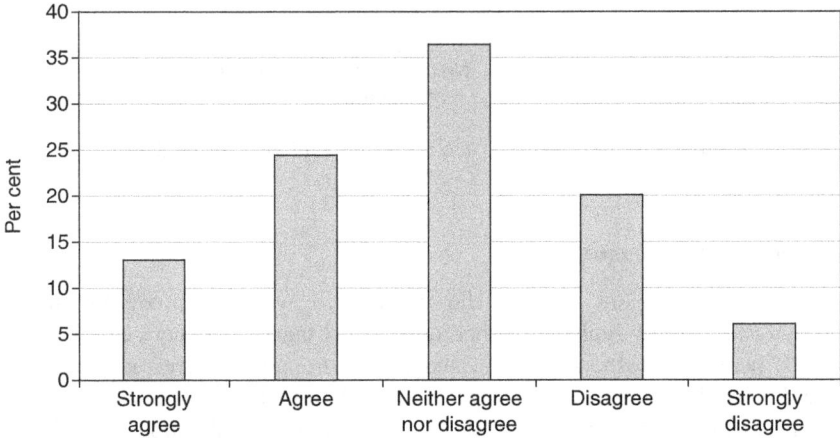

Figure 5.18 Responses to the statement: "The wind farm will bring down the local property prices".

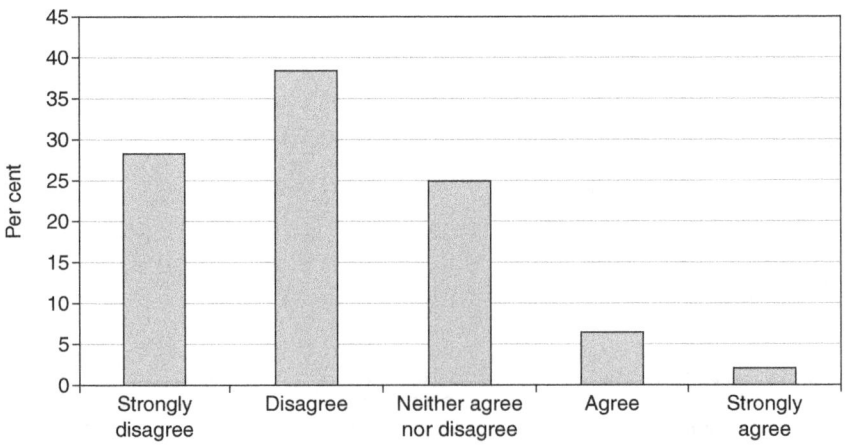

Figure 5.19 Responses to the statement: "The wind farm will attract tourists".

Noise impact

Respondents were asked their opinion about the wind farm being noisy; 38 per cent neither agreed nor disagreed, while 38 per cent either disagreed or strongly disagreed (Figure 5.20).

Cost of electricity

Asked if the wind farm would generate costlier electricity, 35.5 per cent of respondents replied "neither agree nor disagree", although 37 per cent disagreed or strongly disagreed with the statement (Figure 5.21).

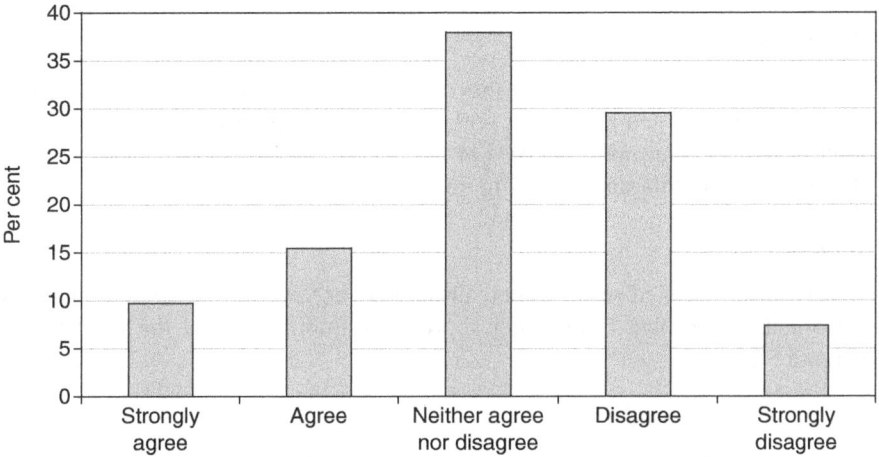

Figure 5.20 Responses to the statement: "The wind farm will be unpleasantly noisy".

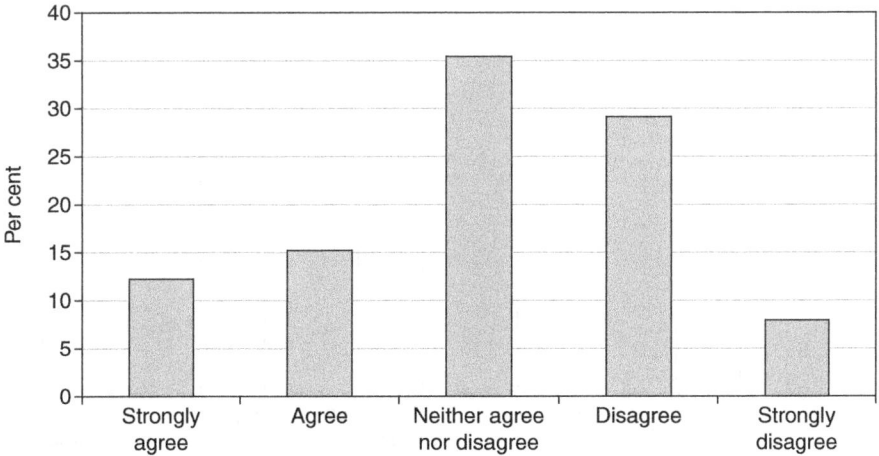

Figure 5.21 Responses to the statement: "The wind farm will generate costlier electricity than if it was generated by ordinary fuel".

Dependency on foreign fuels

Another question asked whether "The wind farm will help to free the country from dependence on foreign fuel"; 40.5 per cent of respondents agreed with the statement (Figure 5.22).

Contextual factors

The contextual factors surveyed were limited to: trust and procedural fairness; the information provided: the co-operative model and its perceived advantages and disadvantages; and respondents' availability to join a co-operative scheme and their reasons for doing so.

Trust and procedural fairness

Respondents were asked whether they, "Trust the developers of the wind farm in the way they deal and have dealt with the local community". Most respondents replied "neither agree nor disagree" (44.5 per cent), although more people disagreed than agreed with this statement (Figure 5.23).

Information

Nearly 60 per cent of respondents "disagreed" or "strongly disagreed" with the statement, "I feel that I have been thoroughly informed about the wind farm". (Figure 5.24).

The co-operative model

Items 13 to 13.7 of the questionnaire (see Appendix A) surveyed respondents' opinions regarding the wind farm co-operative model. This second part of the

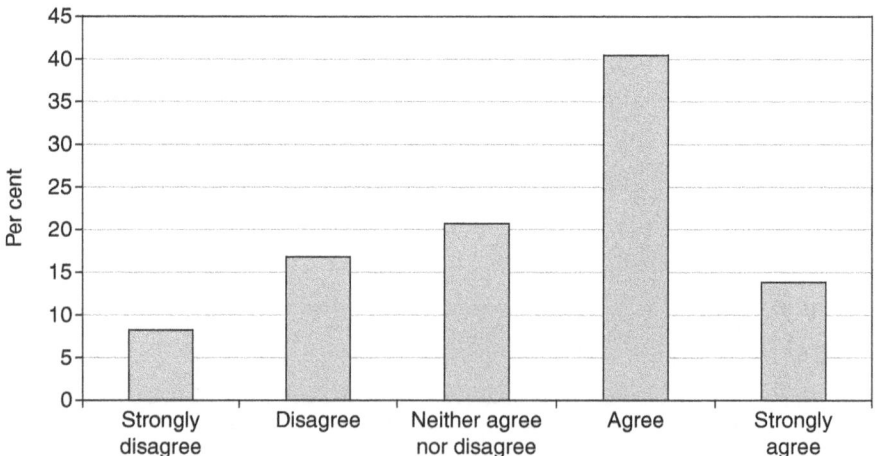

Figure 5.22 Responses to the statement: "The wind farm will help to free the country from dependence on foreign fuels".

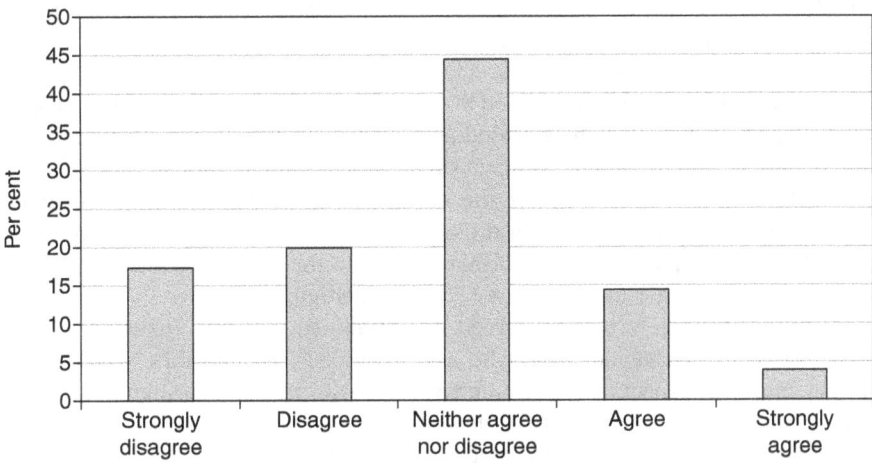

Figure 5.23 Responses to the statement: "I trust the developers of the wind farm in the way they deal and have dealt with the local community".

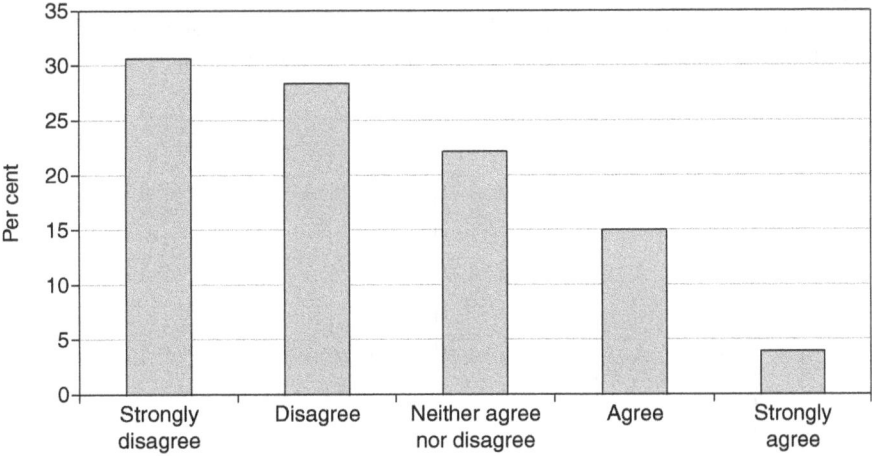

Figure 5.24 Responses to the statement: "I feel that I have been thoroughly informed about the wind farm".

questionnaire received fewer replies than the first part. It is possible that some respondents were not confident in their ability to state an opinion regarding either the proposed wind farm co-operative or a hypothetical wind farm co-operative scheme. Even in the cases of sites where a co-operative scheme was offered, respondents claimed not to be aware of it; very few might, therefore, have been familiar with the idea of a wind farm co-operative.

When respondents were asked if they agreed with the statement, "The co-operative will be just a ploy to buy residents' consensus", 32.5 per cent replied "neither agree nor disagree" while the combined "agree" and "strongly agree" responses totalled 44 per cent (Figure 5.25).

Respondents largely acknowledged that the co-operative would give local people an opportunity to benefit from the wind farm's revenue: 48 per cent of respondents agreed with the statement and 3 per cent strongly agreed (Figure 5.26).

Residents were asked whether the co-operative would create a permanent divide in the community between those who joined and those who opposed it, 38.5 per cent neither agreed nor disagreed, while the combined share of those who agreed or strongly agreed was 42.5 per cent (Figure 5.27).

Respondents largely neither agreed nor disagreed with the statement, "The co-operative will persuade those who are undecided to support the wind farm" (44 per cent). In respect of the remainder, there was a slight prevalence of those who either disagreed or strongly disagreed (30 per cent) compared to those who agreed or strongly agreed (26 per cent) (Figure 5.28).

When asked about the statement that the co-operative "would offer the worst compensation for those who oppose the wind farm because their decision not to join means they would not receive any revenue", 41.5 per cent neither agreed nor disagreed, but 38 per cent agreed or strongly agreed (Figure 5.29).

Respondents were also asked whether they agreed with a statement that the wind farm co-operative "would involve local people not only financially but also

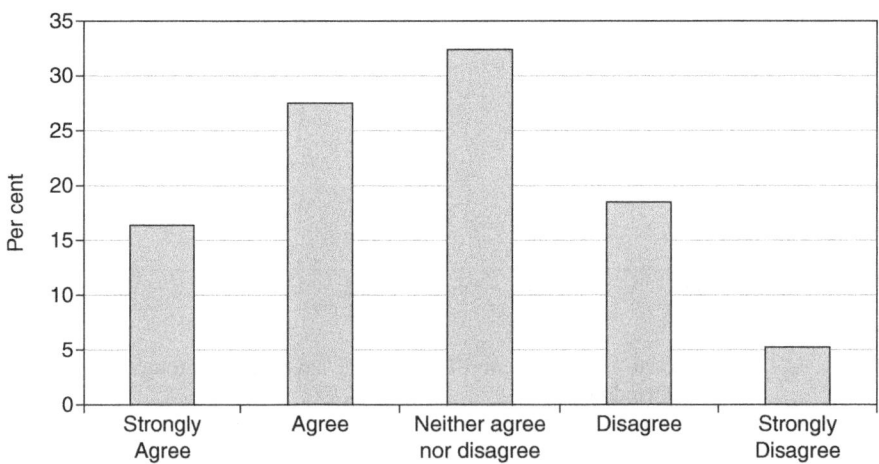

Figure 5.25 Responses to the statement: "The co-operative will just be a ploy to buy residents' consensus".

Note

The statement used varied slightly between co-operative and non-co-operative scheme sites. For co-operative scheme sites, the statement was as indicated above, while for commercial scheme sites the statement "The co-operative would just be a ploy to buy residents' consensus" was used.

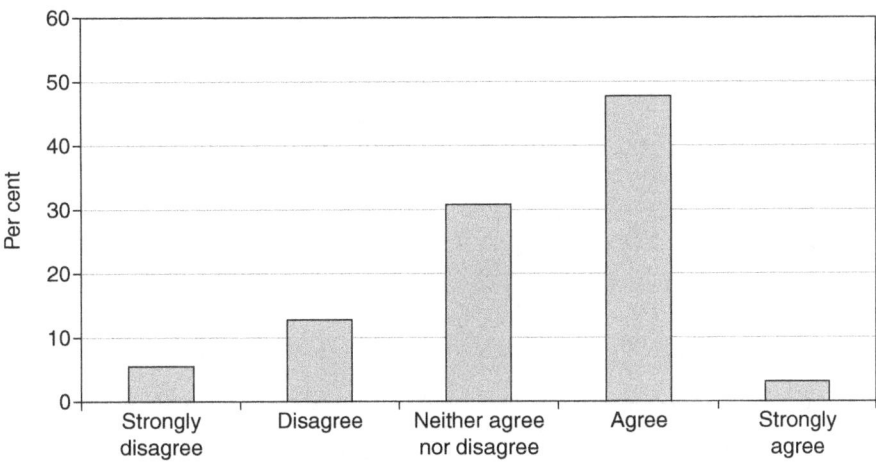

Figure 5.26 Responses to the statement: "The co-operative will give locals the chance to benefit from the revenue of the wind farm".

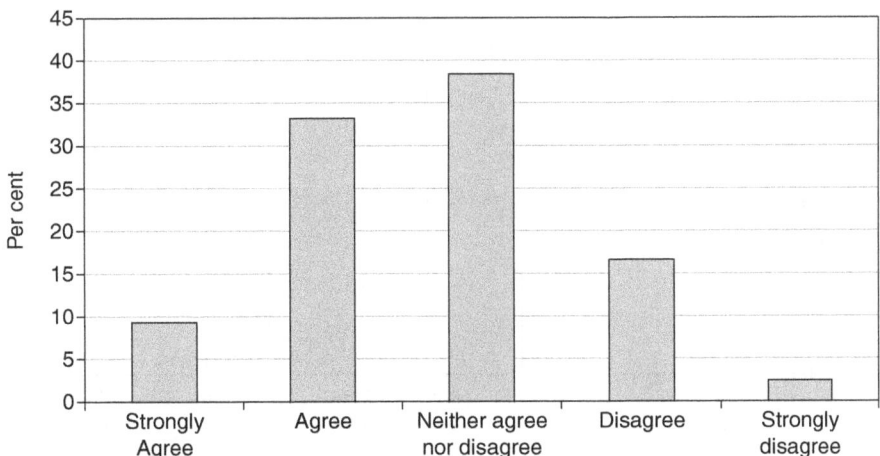

Figure 5.27 Responses to the statement: "The co-operative will create a permanent divide in the local community between those who would join and those who would oppose the wind farm".

in its management: it would create a stable network of local residents who might support further community activities and projects". Thirty-six per cent agreed with the statement, and 5 per cent strongly agreed, while 34.5 per cent neither agreed nor disagreed (Figure 5.30).

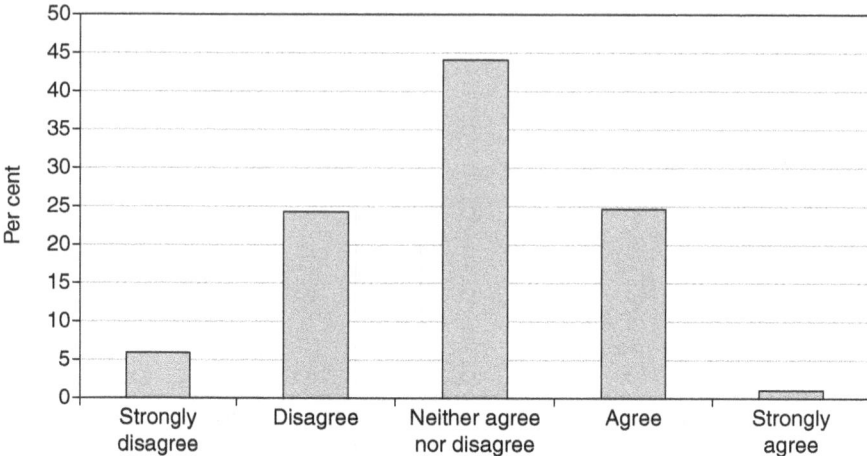

Figure 5.28 Responses to the statement: "The co-operative will persuade those who are undecided to support the wind farm".

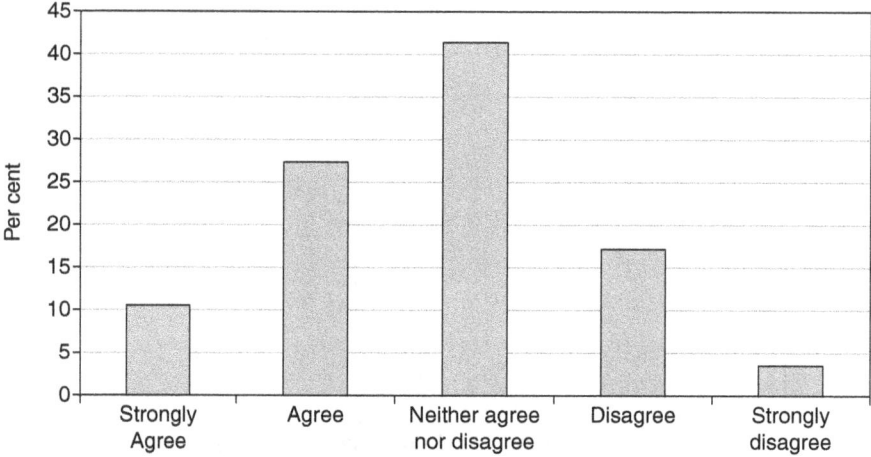

Figure 5.29 Responses to the statement: "The co-operative will offer the worst compensation for those who oppose the wind farm because their decision not to join means they will not receive any revenue".

The statement "The co-operative will persuade even opponents of the wind farm to accept the development" obtained 39 per cent of neither agree nor disagree replies, while 43.5 per cent said they disagreed or strongly disagreed (Figure 5.31).

When they were asked to give their opinion on the statement that the co-operative scheme would not make any difference in terms of support, 54 per cent of respondents agreed and 11.5 per cent strongly agreed (Figure 5.32).

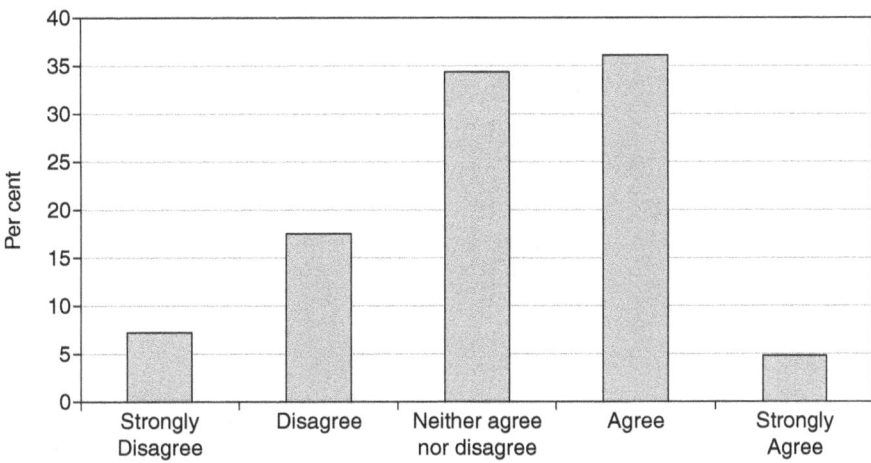

Figure 5.30 Responses to the statement: "The co-operative will involve local people not only financially but also in its management: it will create a stable network of local residents who might support further community activities and projects".

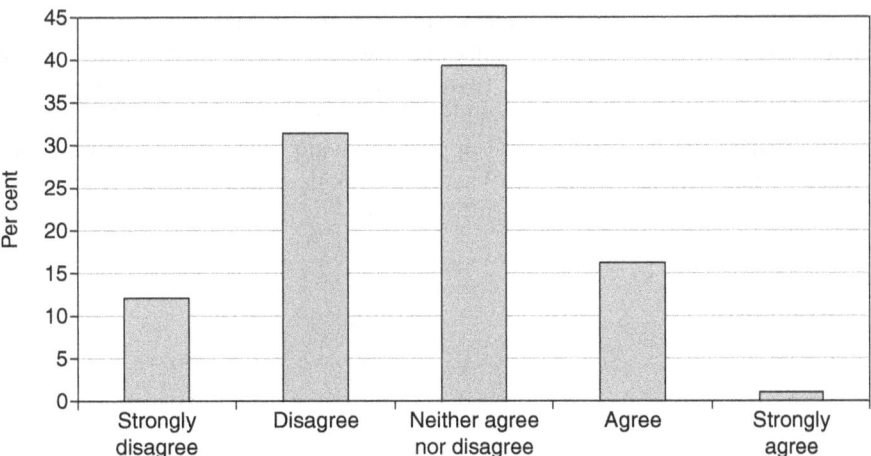

Figure 5.31 Responses to the statement: "The co-operative will persuade even opponents of the wind farm to accept the development".

Respondents were then asked whether they would invest in a co-operative scheme if the price of a single share was £250; 55 per cent answered "No", while 45 per cent said "Yes". Nevertheless, the number that did not respond to this question was relatively high (13.3 per cent); if these are included alongside the "Yes" and "No" responses the latter become, respectively, 47 per cent and 39 per cent.

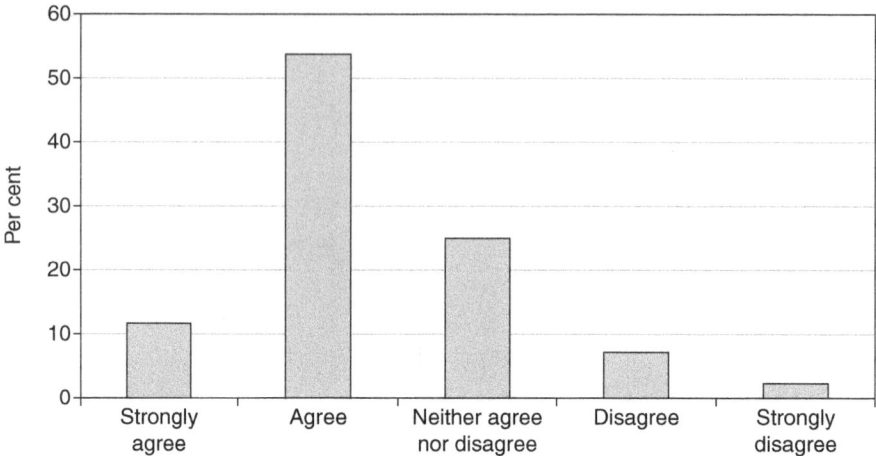

Figure 5.32 Responses to the statement: "The co-operative will not make any difference. People will support or oppose the wind farm regardless of whether there is a co-operative scheme or not".

Respondents were also asked about the reasons for their decision to invest or not to invest. The statement, "I think that it would be a good investment opportunity" obtained 36 per cent of "agree" and 3.5 per cent of "strongly agree" replies, while 33.5 per cent said "neither agree nor disagree" (figure 5.33).

In response to the statement "I oppose the wind farm so I would never join in", 37.5 per cent disagreed and 19 per cent strongly disagreed (Figure 5.34).

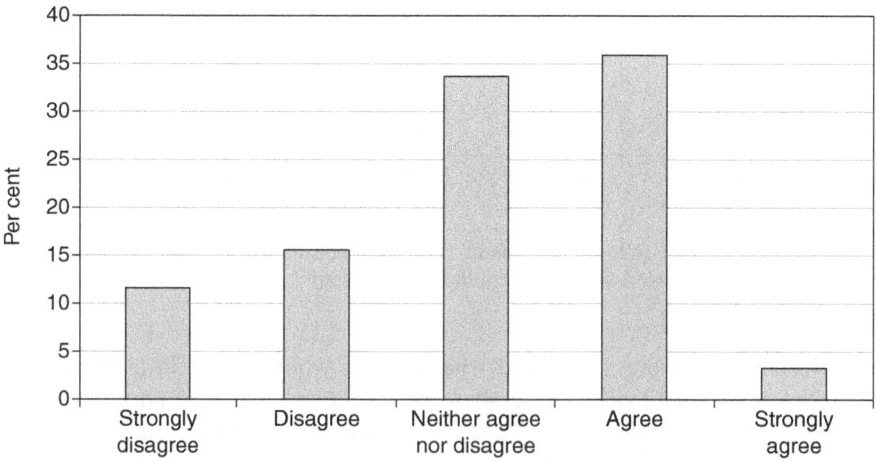

Figure 5.33 Responses to the statement: "I think that it would be a good investment opportunity".

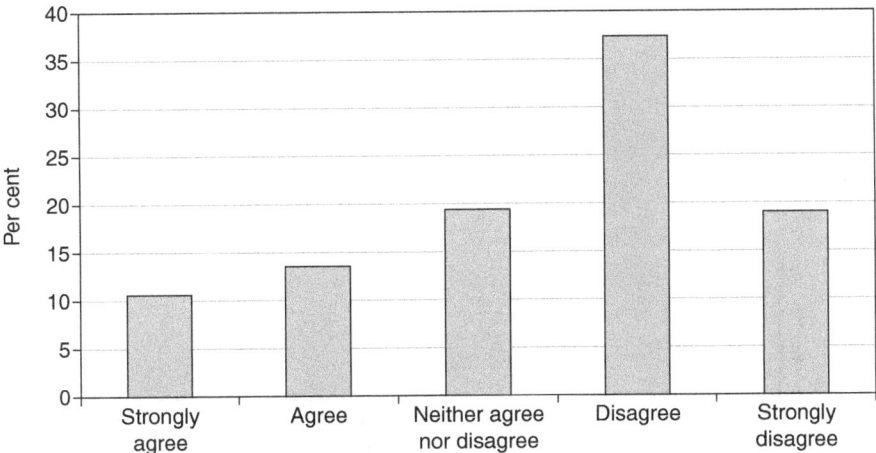

Figure 5.34 Responses to the statement: "I oppose the wind farm so I would never join in".

The statement "I believe that we all should do something to fight climate change, therefore I would join" received support from 46 per cent of respondents who either agreed or strongly agreed (Figure 5.35), suggesting that global environmental concern might be a prime motivator for joining a co-operative scheme.

When evaluating the statement "I couldn't afford to buy the shares", 39 per cent of respondents disagreed and 11 per cent strongly disagreed (Figure 5.36), although about 20 per cent agreed or strongly agreed.

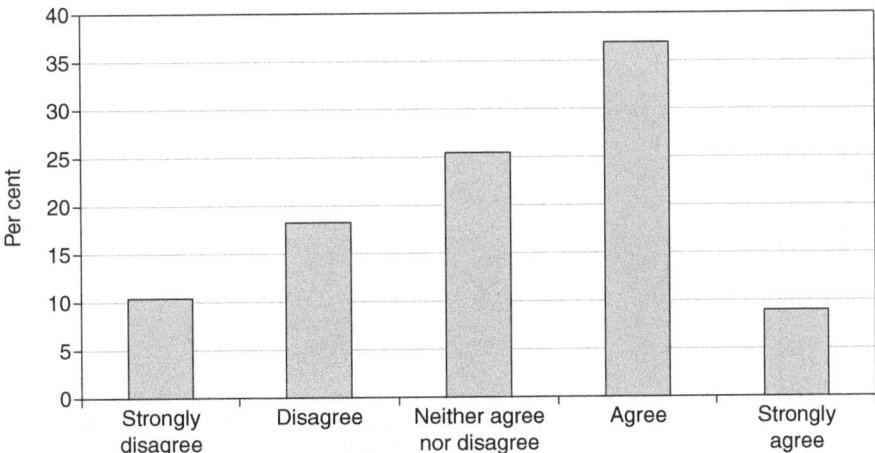

Figure 5.35 Responses to the statement: "I believe that we all should do something to fight climate change, therefore I would join".

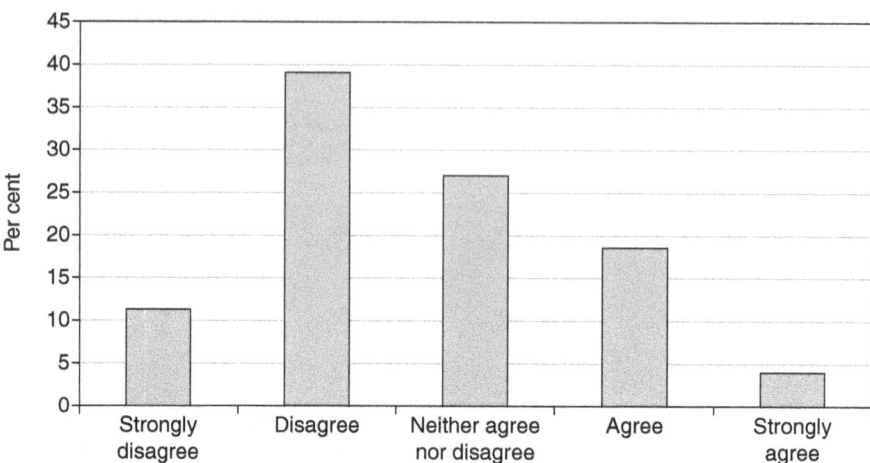

Figure 5.36 Responses to the statement: "I couldn't afford to buy shares".

Respondents were asked whether they agreed or disagreed with the statement, "If people around me, in my community, would support it, so would I"; 36.5 per cent neither agreed nor disagreed, but 47.5 per cent disagreed or strongly disagreed (Figure 5.37).

The statement "I don't care about the wind farm and so I would not care about the co-operative" received 45.5 per cent of responses that 'disagree' and 16.5 per cent that 'strongly disagree' (Figure 5.38).

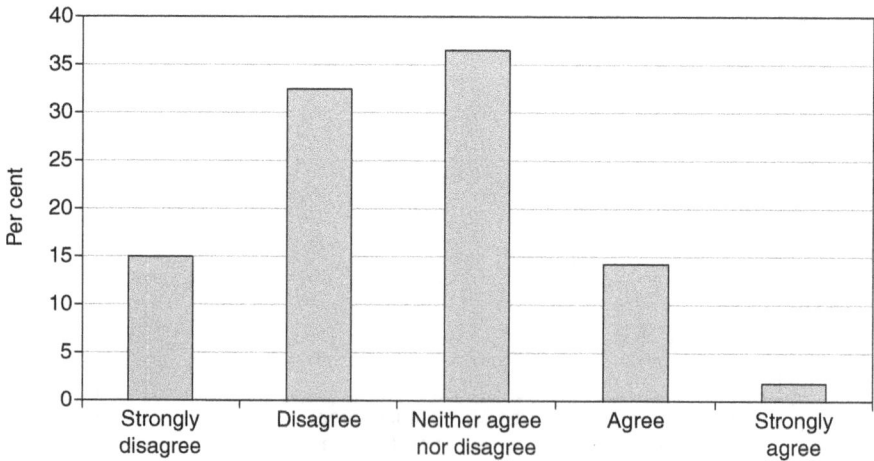

Figure 5.37 Responses to the statement: "If people around me, in my community, would support it, so would I".

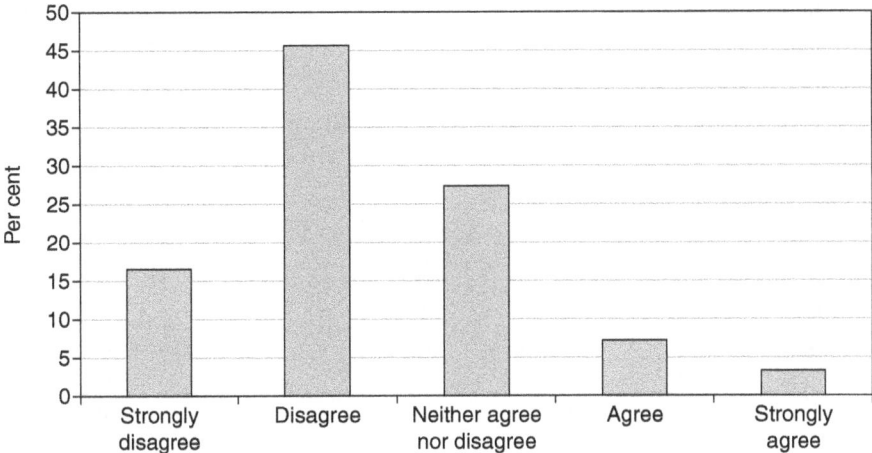

Figure 5.38 Responses to the statement: "I don't care about the wind farm and so I would not care about the co-operative".

In response to the statement, "I would be able to have a say in the development of the wind farm and its management", 36.5 per cent of respondents answered "neither agree nor disagree" while 37.5 per cent said either "agree" or "strongly agree" (Figure 5.39).

Respondents were then asked, "All in all, what do you think of a 'community wind farm co-operative'?"; 41.5 per cent agreed that "It is a good idea" while 10.5 per cent agreed that "It is an excellent idea" (Figure 5.40).

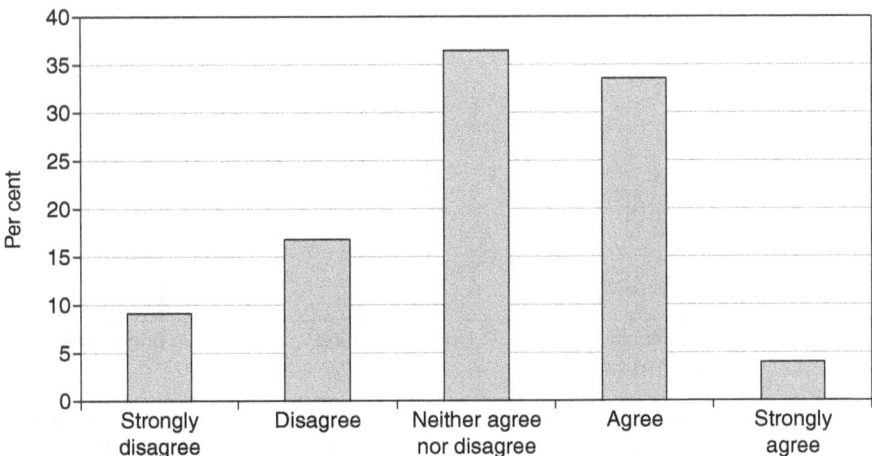

Figure 5.39 Responses to the statement: "I would be able to have a say in the development of the wind farm and its management".

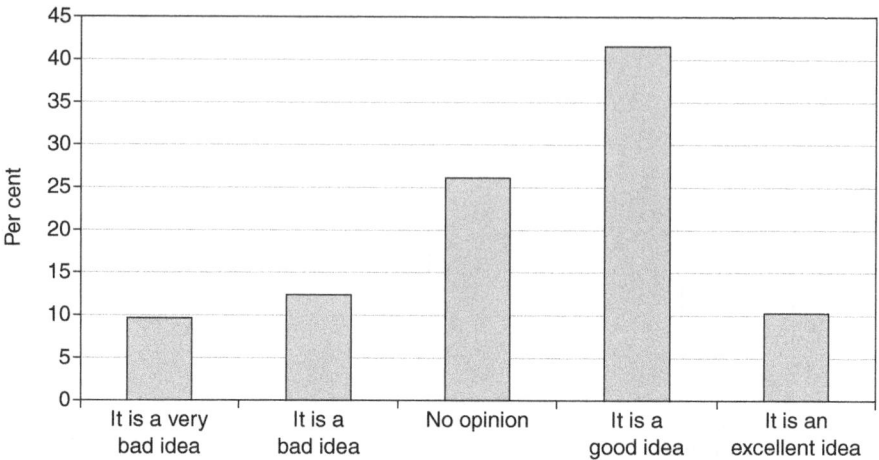

Figure 5.40 Responses to the question: "All in all, what do you think of a 'community wind farm co-operative'?".

Respondents were finally asked whether they would prefer a co-operative scheme or a community fund scheme; 38.5 per cent expressed a preference for the co-operative scheme, while 36 per cent answered "no preference" and 26 per cent said they would prefer the community fund (Figure 5.41).

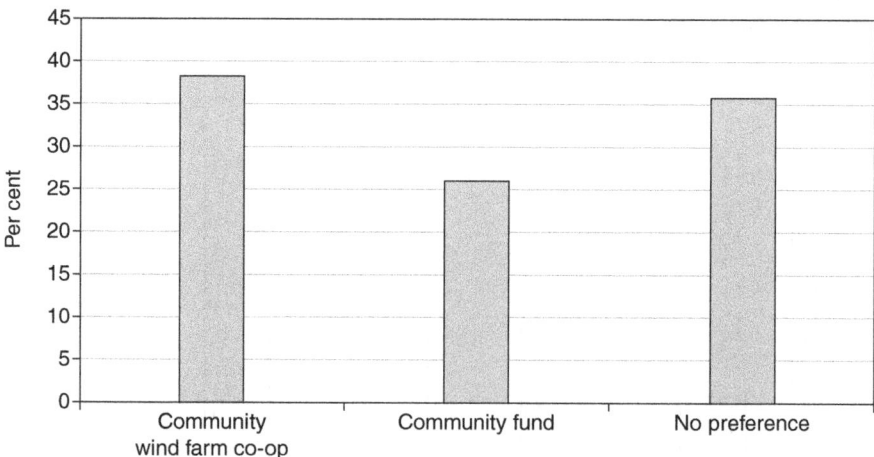

Figure 5.41 Responses to the question: "If you could choose ONE of the schemes that we have presented for your local wind farm, which one would you choose?".

Bivariate correlations

Statistical tests used to assess correlations

Following Bryman and Cramer (2009), different methods were used to assess bivariate correlation, depending on the type of variable analysed. If one of the variables was nominal, a cross tabulation associated to a chi-square test and a Cramer's V test were employed to assess the significance and strength of the relationship. If both variables were ordinal, Kendall's tau was used. Kendall's tau is a more solid test than the alternative, Spearman's rho, and possibly more accurate in small data sets with a large number of tied ranks (Field, 2009).

In each test run using SPSS (Statistical Package for the Social Sciences), one of the two variables involved is the item named "opinionwf" in the SPSS environment. This item corresponds to question 7 in the questionnaire: "What do you think of the presence of this wind farm in your area?" (see Appendix A). A multiple-choice answer was provided in the form of a 5-point Likert scale, ranging from "strongly disagree" to "strongly agree". Such a variable, despite being treated by some researchers as "interval", is more accurately defined as "ordinal", because it is not possible to be certain that the respondents would perceive the difference between, say, "strongly agree" and "agree" as equal to that between "agree" and "neither agree nor disagree" (Bryman and Cramer, 2009; Field, 2009).

The implications of "opinionwf" being ordinal are that parametric tests are ruled out, even though the matter is not agreed between scholars and some nonetheless argue in favour of using parametric tests (Bryman and Cramer, 2009; Field, 2009). In this case, non-parametric tests were used, thereby following the mainstream stance that parametric tests can only be used if certain assumptions, such as the data being measured through interval variables, are met (Field, 2009).

In order to provide the reader with a complete view of results independently from their significance, all are reported. The order of the tests follows the order of the items used to measure the variables in the questionnaire.

Personal resources and demographic variables

This section presents the tests of correlation regarding the variables pertaining to the group presented in Chapter 2, which were included in the questionnaire.

Location

As demonstrated in Tables 5.10–5.12, there is a significant relationship at the 0.05 level, i.e. $p > 0.5$, between respondents' location and their opinions about the proposed wind farm. However, 5 per cent of cells have an expected count that is less than five and less than the minimum expected count of 4.13; therefore, the test should be replicated after collecting more data to confirm the association (Field, 2009). An intervening variable (or variables) must explain the variance in opinion. The relationship is anyway very weak, as the Cramer's V coefficient, which always ranges between 0 and 1, shows.

Table 5.10 Cross tabulation between the variables "opinion about the locally proposed wind farm" and "name of the wind farm"

			Name of wind farm				Total
			Cushnie	Nigg Hill	Meikle Carewe	Bracco	
Opinion about locally proposed wind farm	Strongly disagree	Count	23	23	13	3	62
		% within opinion about locally proposed wind farm	37.1%	37.1%	21.0%	4.8%	100.0%
		% within name of wind farm	20.5%	28.0%	17.8%	7.3%	20.1%
	Disagree	Count	16	5	7	3	31
		% within opinion about locally proposed wind farm	51.6%	16.1%	22.6%	9.7%	100.0%
		% within name of wind farm	14.3%	6.1%	9.6%	7.3%	10.1%
	Neither agree nor disagree	Count	22	19	23	19	83
		% within opinion about locally proposed wind farm	26.5%	22.9%	27.7%	22.9%	100.0%
		% within name of wind farm	19.6%	23.2%	31.5%	46.3%	26.9%
	Agree	Count	33	21	24	9	87
		% within opinion about locally proposed wind farm	37.9%	24.1%	27.6%	10.3%	100.0%
		% within name of wind farm	29.5%	25.6%	32.9%	22.0%	28.2%
	Strongly agree	Count	18	14	6	7	45
		% within opinion about locally proposed wind farm	40.0%	31.1%	13.3%	15.6%	100.0%
		% within name of wind farm	16.1%	17.1%	8.2%	17.1%	14.6%
Total		Count	112	82	73	41	308
		% within opinion about locally proposed wind farm	36.4%	26.6%	23.7%	13.3%	100.0%
		% within name of wind farm	100.0%	100.0%	100.0%	100.0%	100.0%

Table 5.11 Chi-square tests

	Value	Df	Asymp. sig. (2-sided)
Pearson chi-square	22.734[a]	12	0.030
Likelihood ratio	23.149	12	0.026
Linear-by-linear association	0.720	1	0.396
N of valid cases	308	–	–

Notes
a: 1 cell (5.0%) has an expected count below 5. The minimum expected count is 4.13.
Df: degrees of freedom.
Asymp. sig.: asymptotic significance.

Table 5.12 Symmetric measures

		Value	Approx. sig.
Nominal by nominal	Phi	0.272	0.030
	Cramer's V	0.157	0.030
N of valid cases		308	–

Opinion about the wind farm and distance of respondents

Contrary to expectations, no significant relationship was found between respondents' distance from the proposed wind farm and their opinion of it (Table 5.13).

Deprivation vs affluence

When the proposed wind farm sites were simply divided into deprived and not deprived, a significant relationship could not be found between the social context of deprivation/affluence and opinions about the proposed wind farm (Table 5.14).

Table 5.13 Correlations between "residents' distance from the proposed site" and "opinion about the locally proposed wind farm"

			Opinion about the locally proposed wind farm
Kendall's tau_b	*Residents' distance from the proposed site*	Correlation coefficient	0.057
		Sig. (1-tailed)	0.134
		N	308

Table 5.14 Correlations between "opinion about the proposed wind farm" and "social context deprived/affluent"

			Opinion about the locally proposed wind farm
Kendall's tau_b	Social context deprived/affluent	Correlation coefficient	0.012
		Sig. (1-tailed)	0.407
		N	308

A second deprivation variable was created, called "socialcontext2", recoding the wind farm name variable which placed the names in order according to the surveyed area's average SIMD. "1" corresponded to the least deprived (Meikle Carewe) and "4" to the most deprived (Bracco). Once again, the result did not show any significant relationship between affluence and opinions about the wind farm (Table 5.15).

The relationship between deprivation and opinion about the wind farm, i.e. the presence of an opinion, was then assessed using a cross tabulation and its relative chi-square and Cramer's V between the deprivation variable "socialcontext2", described above, and a new variable called "presence of opinion", obtained by recoding the variable of opinion about the wind farm. The recoding process entailed generating a dichotomous variable with scores 1, opinion, and 2, no opinion, obtained from the 5-point Likert scale values of the original variable ("strongly disagree", etc.). In this case, a significant correlation was found at the 0.01 level, $p>0.01$, albeit of modest strength (Tables 5.16–5.18). This suggests that the more a context is deprived, the more citizens are apathetic towards proposed developments in their area.

Level of education

The correlation between respondents' educational level and opinion about the wind farm was then tested. No significant relationship was found (Table 5.19).

Table 5.15 Correlations between "opinion about the proposed wind farm" and "socialcontext2"

			Socialcontext2
Kendall's tau_b	Opinion about the locally proposed wind farm	Correlation coefficient	0.026
		Sig. (1-tailed)	0.295
		N	308

Table 5.16 Cross tabulation between the variables "presence of opinion" and "socialcontext2"

			Socialcontext2				Total
			1.00	*2.00*	*3.00*	*4.00*	
Presence of opinion	*Opinion*	Count	50	90	63	22	225
		% within presence of opinion	22.2%	40.0%	28.0%	9.8%	100.0%
		% within socialcontext2	68.5%	80.4%	76.8%	53.7%	73.1%
	No opinion	Count	23	22	19	19	83
		% within presence of opinion	27.7%	26.5%	22.9%	22.9%	100.0%
		% within socialcontext2	31.5%	19.6%	23.2%	46.3%	26.9%
Total		Count	73	112	82	41	308
		% within presence of opinion	23.7%	36.4%	26.6%	13.3%	100.0%

Table 5.17 Chi-square tests

	Value	*Df*	*Asymp. sig. (2-sided)*
Pearson chi-square	12.234[a]	3	0.007
Likelihood ratio	11.628	3	0.009
Linear-by-linear association	1.557	1	0.212
N of valid cases	308	–	–

Note
a: no cells (0.0%) have an expected count below 5. The minimum expected count is 11.05.
Df: degrees of freedom.
Asymp. sig: asymptotic significance.

Table 5.18 Symmetric measures

		Value	Approx. sig.
Nominal by nominal	Phi	0.199	0.007
	Cramer's V	0.199	0.007
N of valid cases		308	–

Table 5.19 Correlations between "opinion about the proposed wind farm" and "level of education"

			Opinion about the locally proposed wind farm
Kendall's tau_b	Level of education	Correlation coefficient	–0.075
		Sig. (1-tailed)	0.061
		N	297

Number of household members

Looking at the relationship between household size and opinion about the proposed wind farm, a significant weak correlation was found (p<0.05), meaning that larger households were slightly more supportive (Table 5.20).

Income

No significant correlation was found between household income and opinion about the proposed wind farm (Table 5.21).

When the correlation between income per family member and opinion about the proposed wind farm was checked, a weak negative significant correlation (p<0.05) was again found. This means that, in the sample, those with a higher income per family member were less likely to approve of the proposed wind farm (Table 5.22).

Table 5.20 Correlations between "opinion about the proposed wind farm" and "number of household members"

			Opinion about the locally proposed wind farm
Kendall's tau_b	Number of household members	Correlation coefficient	0.114*
		Sig. (1-tailed)	0.013
		N	263

Table 5.21 Correlations between "opinion about the proposed wind farm" and "household income"

			Opinion about the locally proposed wind farm
Kendall's tau_b	*Household income*	Correlation coefficient	−0.067
		Sig. (1-tailed)	0.084
		N	271

Table 5.22 Correlations between "opinion about the proposed wind farm" and "estimation of income per family member"

			Opinion about the locally proposed wind farm
Kendall's tau_b	*Estimation of income per family member*	Correlation coefficient	−0.111
		Sig. (1-tailed)	0.014
		N	233

Knowledge about wind energy

A range of questions assessed respondents' knowledge about wind energy, specifically: comparable pollution caused by wind energy production (wind vs coal), comparable cost of wind energy (wind vs coal), the intermittency of wind generation from a single generator, and the classification of wind energy as a renewable energy source (questions 4.1–4.4, see Appendix A).

The rationale, as explained earlier, was to assess how knowledge influenced levels of consent regarding a locally proposed wind farm.

Therefore, these variables were computed to create an index, named "knowwind", which was built for any respondent who had answered at least three questions out of four, following the procedure described by Bryman and Cramer (2009).

A correlation test between this index and opinions regarding the wind farm was generated, returning a result on the border of the 0.05 significance level (Table 5.23).

Repeating the Kendall's tau test after recoding the variable regarding the opinion about the wind farm could return a significant result. In fact, the formula for calculating Kendall's tau appears as follows (Nelsen, 2011):

$$\tau = \frac{(\text{number of concordant pairs}) - (\text{number of discordant pairs})}{\frac{1}{2}n(n-1)}$$

Therefore, it is possible to assume that increasing the number of concordant pairs while reducing the number of discordant pairs will eventually increase the value of tau statistic and its significance.

Table 5.23 Correlations between "opinion about the proposed wind farm" and "knowwind"

			Opinion about the locally proposed wind farm
Kendall's tau_b	*Knowwind*	Correlation coefficient	−0.078
		Sig. (1-tailed)	0.051
		N	303

This could be done by recoding the 5-point Likert scale answers of the item regarding the opinion about the proposed wind farm, named "opinionwf", in a 3-point scale ("disagree", "don't know", "agree"), named "opinionwf3items". Testing again for a correlation, the coefficient was found to be larger and significant at the 0.05 level (Table 5.24), therefore showing that the higher the level of knowledge, the less respondents agree with the proposed wind farm.

Awareness of the proposed wind farm

Looking at the correlation between awareness of the proposed wind farm and opinion about the wind farm, a significant correlation ($p<0.0005$) was found that presents a Cramer's V value of 0.35 (Tables 5.25–5.27).

Seeing the proposed wind farm site from home

The correlation between seeing the proposed wind farm site from home and opinion about the wind farm was also significant ($p<0.0005$). Here, to limit the size of the cross tabulation, the variable "opinionwf3items" was used instead of "opinionwf" (Tables 5.28–5.30).

Frequency of view of the proposed site

The correlation between how often respondents would see the proposed site, in cases where they were not able to see it from their home, and opinions about the proposed wind farm was then assessed.

The relationship was negative, which means that those who answered "never" or "rarely" were more likely to support the wind farm. Nevertheless, the significance level was not achieved (Table 5.31). A significant relationship

Table 5.24 Correlations between "opinionwf3items" and "knowwind"

			Opinionwf3items
Kendall's tau_b	*Knowwind*	Correlation coefficient	−0.096
		Sig. (1-tailed)	0.028
		N	303

Table 5.25 Cross tabulation between the variables "opinion about the locally proposed wind farm" and "awareness of the proposed wind farm"

			Awareness of proposed wind farm		Total
			No	Yes	
Opinion about locally proposed wind farm	Strongly disagree	Count	7	54	61
		% within opinion about locally proposed wind farm	11.5%	88.5%	100.0%
		% within awareness of proposed wind farm	7.1%	26.0%	19.9%
	Disagree	Count	9	22	31
		% within opinion about locally proposed wind farm	29.0%	71.0%	100.0%
		% within awareness of proposed wind farm	9.2%	10.6%	10.1%
	Neither agree nor disagree	Count	47	36	83
		% within opinion about locally proposed wind farm	56.6%	43.4%	100.0%
		% within awareness of proposed wind farm	48.0%	17.3%	27.1%
	Agree	Count	26	60	86
		% within opinion about locally proposed wind farm	30.2%	69.8%	100.0%
		% within awareness of proposed wind farm	26.5%	28.8%	28.1%
	Strongly agree	Count	9	36	45
		% within opinion about locally proposed wind farm	20.0%	80.0%	100.0%
		% within awareness of proposed wind farm	9.2%	17.3%	14.7%
Total		Count	98	208	306
		% within opinion about locally proposed wind farm	32.0%	68.0%	100.0%

Table 5.26 Chi-square tests

	Value	Df	Asymp. sig. (2-sided)
Pearson chi-square	38.152[a]	4	0.000
Likelihood ratio	38.898	4	0.000
Linear-by-linear association	1.578	1	0.209
N of valid cases	306	–	–

Notes
a: no cells (0.0%) have an expected count below 5. The minimum expected count is 9.93.
Df: degrees of freedom.
Asymp. sig.: asymptotic significance.

Table 5.27 Symmetric measures

		Value	Approx. sig.
Nominal by nominal	Phi	0.353	0.000
	Cramer's V	0.353	0.000
N of valid cases		306	–

was, however, achieved when the correlation was tested again using the reduced variable of opinion about the wind farm, "opinionwf3items" (p<0.05) (Table 5.32).

Awareness of the presence of a community scheme

The following section examines respondents' awareness of the presence of a community scheme. As was shown earlier, the vast majority of respondents were not aware (see Table 5.5).

The variable is nominal and therefore looking for a correlation with "opinion about the wind farm" entailed generating a cross tabulation and calculating the Chi-Square, but this statistic requires the presence of a minimum number of cases within the cross tabulation, which could not be met without collapsing both variables. So, in the case of opinion about the wind farm, the "opinionwf3items" variable was used, which transforms the original variable from a 5-point Likert scale into a 3-point scale. In the case of the nominal variable about the awareness of the schemes, this was repeatedly collapsed until the cross tabulation resulted in a table bearing no zeros in any box. This was achieved only when the variable named "awarescheme3" was ultimately reduced to a dichotomous variable with two answer modes: "any community scheme" and "not aware of any" (Table 5.33).

Table 5.28 Cross tabulation between the variables "opinionwf3items" and "seeing the wind farm site from home"

			Seeing the wind farm site from home			Total
			No	Yes	I don't know where it is	
Opinionwf3items	Disagree + strongly disagree	Count	34	48	11	93
		% within opinionwf3items	36.6%	51.6%	11.8%	100.0%
		% within seeing wind farm site from home	24.1%	51.1%	15.1%	30.2%
	Neither agree nor disagree	Count	34	13	36	83
		% within opinionwf3items	41.0%	15.7%	43.4%	100.0%
		% within seeing wind farm site from home	24.1%	13.8%	49.3%	26.9%
	Agree + strongly agree	Count	73	33	26	132
		% within opinionwf3items	55.3%	25.0%	19.7%	100.0%
		% within seeing wind farm site from home	51.8%	35.1%	35.6%	42.9%
Total		Count	141	94	73	308
		% within opinionwf3items	45.8%	30.5%	23.7%	100.0%

Table 5.29 Chi-square tests

	Value	Df	Asymp. sig. (2-sided)
Pearson chi-square	45.618[a]	4	0.000
Likelihood ratio	43.268	4	0.000
Linear-by-linear association	1.660	1	0.198
N of valid cases	308	–	–

Notes
a: no cells (0.0%) have an expected count below 5. The minimum expected count is 19.67.
Df: degrees of freedom.
Asymp. sig.: asymptotic significance.

Table 5.30 Symmetric measures

		Value	Approx. sig.
Nominal by	Phi	0.385	0.000
nominal	Cramer's V	0.272	0.000
N of valid cases		308	–

Table 5.31 Correlations between "opinion about the proposed wind farm" and "how often see the wind farm site"

			Opinion about the locally proposed wind farm
Kendall's tau_b	*How often see the wind farm site*	Correlation coefficient	–0.093
		Sig. (1-tailed)	0.091
		N	140

Table 5.32 Correlations between "opinionwf3items" and "how often see the wind farm site"

			Opinionwf3items
Kendall's tau_b	*How often see the wind farm site*	Correlation coefficient	–0.123
		Sig. (1-tailed)	0.047
		N	140

The correlation is significant at the 0.01 level, although the Cramer's V coefficient of 0.219 suggests a weak correlation, meaning that those more aware of any scheme tended to be less in favour (Tables 5.34 and 5.35).

Table 5.33 Cross tabulation between the variables "opinionwf3items" and "awarescheme3"

			Awarescheme3		Total
			Any community scheme	*Not aware of any*	
Opinion-wf3items	Disagree + strongly disagree	Count	21	70	91
		% within opinion-wf3items	23.1%	76.9%	100.0%
		% within awarescheme3	53.8%	26.8%	30.3%
	Neither agree nor disagree	Count	3	77	80
		% within opinion-wf3items	3.8%	96.3%	100.0%
		% within awarescheme3	7.7%	29.5%	26.7%
	Agree + strongly agree	Count	15	114	129
		% within opinion-wf3items	11.6%	88.4%	100.0%
		% within awarescheme3	38.5%	43.7%	43.0%
Total		Count	39	261	300
		% within opinion-wf3items	13.0%	87.0%	100.0%

Table 5.34 Chi-square tests

	Value	*Df*	*Asymp. sig. (2-sided)*
Pearson chi-Square	14.437[a]	2	0.001
Likelihood ratio	15.191	2	0.001
Linear-by-linear association	4.901	1	0.027
N of valid cases	300	–	–

Notes
a: no cells (0.0%) have an expected count below 5. The minimum expected count is 10.40.
Df: degrees of freedom.
Asymp. sig.: asymptotic significance.

Table 5.35 Symmetric measures

		Value	Approx. sig.
Nominal by nominal	Phi	0.219	0.001
	Cramer's V	0.219	0.001
N of valid cases		300	–

Attitudinal factors

In this section, correlations concerning the attitudinal factors, whose frequencies were presented earlier in the chapter, are illustrated.

Local benefits and costs

When testing the relationship between the number of perceived local advantages and disadvantages related to the proposed wind farm and opinion about the wind farm, a significance level of $p < 0.0005$ was achieved (Table 5.36). Hence, subjects who believed that wind farm development would create mostly disadvantages disagreed the most with the project.

Health of the local community

Looking in more detail at the specific range of advantages and disadvantages surveyed, it is possible to see that when individuals believe that the wind farm will harm the health of the local community, they are likely to oppose the development. The relationship is significant at the 0.0005 level (Table 5.37).

Climate change impact

The same level of significance $p < 0.0005$ was found when the belief that the wind farm would help against climate change was tested in relation with opinion about the wind farm (Table 5.38). Respondents who held the belief that the wind farm would help against climate change appeared more supportive of the project.

Table 5.36 Correlations between "opinion about the wind farm" and "local benefits/ disadvantages"

			Opinion about the locally proposed wind farm
Kendall's tau_b	*Local benefits/ disadvantages*	Correlation coefficient	0.684
		Sig. (2-tailed)	0.000
		N	298

Table 5.37 Correlations between "opinion about the wind farm" and "wind farm harms local health"

			Opinion about the locally proposed wind farm
Kendall's tau_b	*Wind farm harms local health*	Correlation coefficient	0.571
		Sig. (2-tailed)	0.000
		N	292

Table 5.38 Correlations between "opinion about the wind farm" and "wind farm helps climate change"

			Opinion about the locally proposed wind farm
Kendall's tau_b	*Wind farm helps climate change*	Correlation coefficient	0.501
		Sig. (2-tailed)	0.000
		N	300

Visual impact

A significant correlation at the 0.0005 level was also found between the belief that the wind farm would look bad on the landscape and opposition to the wind farm, showing that respondents who foresaw a negative visual impact were more likely to have a negative opinion about the wind farm (Table 5.39).

Impact on the local economy

When the correlation between the belief that the wind farm would improve the local economy was tested with opinion about the wind farm, a positive correlation was found at the significance level of 0.0005, meaning that those who held this belief were more supportive of the wind farm (Table 5.40).

Table 5.39 Correlations between "opinion about the wind farm" and "wind farm looks bad on the landscape"

			Opinion about the locally proposed wind farm
Kendall's tau_b	*Wind farm looks bad on the landscape*	Correlation coefficient	0.654
		Sig. (2-tailed)	0.000
		N	295

Table 5.40 Correlations between "opinion about the wind farm" and "wind farm improves the local economy"

			Opinion about the locally proposed wind farm
Kendall's tau_b	*Wind farm improves the local economy*	Correlation coefficient	0.567
		Sig. (2-tailed)	0.000
		N	290

Impact on property prices

Respondents who thought that the wind farm would bring down local property prices were more likely to oppose the wind farm. The correlation held at a significance level of 0.0005 (Table 5.41).

Impact on local tourism

Subjects who disagreed with the statement that the wind farm would attract tourists were more likely to oppose the development, as shown by a correlation significant at the 0.0005 level (Table 5.42).

Noise impact

The belief that the wind farm would be noisy was also significantly correlated with opposition to the wind farm at the 0.0005 level (Table 5.43).

Cost of electricity

A significant positive correlation at the 0.0005 level was also found between the belief that the wind farm would produce costlier electricity and opposition towards the proposed development (Table 5.44).

Table 5.41 Correlations between "opinion about the wind farm" and "wind farm bring down property prices"

			Opinion about the locally proposed wind farm
Kendall's tau_b	*Wind farm bring down property prices*	Correlation coefficient	0.526
		Sig. (2-tailed)	0.000
		N	295

Table 5.42 Correlations between "opinion about the locally proposed wind farm" and "wind farm attracts tourists"

			Opinion about the locally proposed wind farm
Kendall's tau_b	*Wind farm attracts tourists*	Correlation coefficient	0.505
		Sig. (2-tailed)	0.000
		N	294

Table 5.43 Correlations between "opinion about the locally proposed wind farm" and "wind farm is noisy"

			Opinion about the locally proposed wind farm
Kendall's tau_b	*Wind farm is noisy*	Correlation coefficient	0.537
		Sig. (2-tailed)	0.000
		N	295

Table 5.44 Correlations between "opinion about the locally proposed wind farm" and "wind farm generates costlier electricity"

			Opinion about the locally proposed wind farm
Kendall's tau_b	*Wind farm generates costlier electricity*	Correlation coefficient	0.385
		Sig. (2-tailed)	0.000
		N	298

Dependency on foreign fuels

A positive correlation significant at the 0.0005 level was found between the belief that the wind farm would help to tackle the problem of fuel dependency and support for the project (Table 5.45).

The "benefitscostsvalue" scale

A new variable was built computing all the scores which respondents gave to single benefits and costs, or advantages and disadvantages that they were asked about, whose correlations with opinion on the wind farm have been presented. Nine variables used, ranging from effects on health to effects on fuel dependency. The new variable was named "benefitscostsvalue" because it reflects the

Table 5.45 Correlations between "opinion about the locally proposed wind farm" and "wind farm helps with fuel dependency"

			Opinion about the locally proposed wind farm
Kendall's tau_b	*Wind farm helps with fuel dependency*	Correlation coefficient	0.532
		Sig. (2-tailed)	0.000
		N	300

value that each respondent gave to all the specific costs and benefits of the wind farm surveyed.

Once this new variable was tested in relation with opinion about the wind farm, a significant correlation at the 0.0005 level with a "high" coefficient, by the standards indicated by Bryman and Cramer (2009) (Table 5.46), was found.

A hypothesis was made in earlier chapters that personal resources such as education, income and knowledge could influence the perception of benefits and costs regarding the proposed development, and, more generally, about any considered course of pro-environmental action.

In order to assess this hypothesis, the correlation between the variable "benefitscostsvalue" and the "knowwind" value was tested. The relationship proved to be significant at the 0.05 level but very weak and negative, i.e. people who scored higher on knowledge about wind energy tended to see fewer benefits from the project (Table 5.47).

Following the same rationale, the relationship between "benefitscostsvalue" and estimated annual income per family member was tested. The correlation is significant at the 0.05 level but with a very low negative coefficient, meaning that households with a higher income per person tended to see more disadvantages in the project (Table 5.48).

Education was tested with "benefitscostsvalue", but no significant correlation was found (Table 5.49).

Table 5.46 Correlations between "opinion about the locally proposed wind farm" and "benefitscostsvalue"

			Opinion about the locally proposed wind farm
Kendall's tau_b	*benefitscostsvalue*	Correlation coefficient	0.731
		Sig. (2-tailed)	0.000
		N	291

Table 5.47 Correlations between "benefitscostsvalue" and "knowwind"

			Benefitscostsvalue
Kendall's tau_b	*Knowwind*	Correlation coefficient	−0.078
		Sig. (1-tailed)	0.044
		N	289

Table 5.48 Correlations between "benefitscostsvalue" and "estimation of income per family member"

			Benefitscostsvalue
Kendall's tau_b	*Estimation of income per family member*	Correlation coefficient	−0.090
		Sig. (1-tailed)	0.028
		N	226

Table 5.49 Correlations between "benefitscostsvalue" and "level of education"

			Level of education
Kendall's tau_b	*Benefitscostsvalue*	Correlation coefficient	−0.064
		Sig. (1-tailed)	0.079
		N	283

Environmental citizenship

Some items (questions 18–18.3, see Appendix A), looked at the concept of "environmental citizenship", i.e. respondents' sense of responsibility towards either the global, the national or the local environment—or none of them, considering the government, instead, as the only actor responsible for the environment.

The correlation of this variable with opinion about the wind farm was tested looking at the chi-square and Cramer's V coefficient, but no significant correlation was found (Tables 5.50–5.52).

Pro-environmental attitudes vs pro-economy attitudes

The correlation between opinion about the wind farm and the variable that attempted to gauge the pro-environmental attitudes of respondents was examined by asking respondents to choose one of four statements based on a trade-off between environmental protection and economic growth (questions 19–19.3, see Appendix A). This variable appears to have a significant correlation at the 0.05 level with opinion regarding the wind farm, but presents a very low coefficient (Table 5.53).

Table 5.50 Cross tabulation between the variables "opinionwf3items" and "environmental citizenship level"

			Environmental citizenship level				Total
			We all have to do something to protect the global environment	We all have to do something to take care of our country's environment	We all have to do something to protect our local environment	It's a matter for the government, not us, to take care of the environment	
opinionwf3items	Disagree + strongly disagree	Count	32	7	37	6	82
		% within opinionwf3items	39.0%	8.5%	45.1%	7.3%	100.0%
		% within environmental citizenship level	24.1%	23.3%	33.6%	42.9%	28.6%
	Neither agree nor disagree	Count	38	8	29	4	79
		% within opinionwf3items	48.1%	10.1%	36.7%	5.1%	100.0%
		% within environmental citizenship level	28.6%	26.7%	26.4%	28.6%	27.5%
	Agree + strongly agree	Count	63	15	44	4	126
		% within opinionwf3items	50.0%	11.9%	34.9%	3.2%	100.0%
		% within environmental citizenship level	47.4%	50.0%	40.0%	28.6%	43.9%
Total		Count	133	30	110	14	287
		% within opinionwf3items	46.3%	10.5%	38.3%	4.9%	100.0%

Table 5.51 Chi-square tests

	Value	Df	Asymp. sig. (2-sided)
Pearson chi-square	5.093[a]	6	0.532
Likelihood ratio	5.077	6	0.534
Linear-by-linear association	3.934	1	0.047
N of valid cases	287	–	–

Notes
a: two cells (16.7%) have an expected count below 5. The minimum expected count is 3.85.
Df: degrees of freedom.
Asymp. sig.: asymptotic significance.

Table 5.52 Symmetric measures

		Value	Approx. sig.
Nominal by nominal	Phi	0.133	0.532
	Cramer's V	0.094	0.532
N of valid cases		287	–

Table 5.53 Correlations between "opinion about the locally proposed wind farm" and "environment vs economy trade-off"

			Opinion about the locally proposed wind farm
Kendall's tau_b	Environment vs economy trade-off	Correlation coefficient	0.157
		Sig. (2-tailed)	0.002
		N	294

Place attachment

The last two items in the questionnaire (questions 20 and 20.1, see Appendix A) assessed respondents' attachment to the physical and social places surrounding them. Using the single items in the questionnaire, a new scale variable, "Place_attachment2", was created on which respondents ranked between 2 (the maximum level of attachment) and 10 (the minimum level of attachment). The correlation between this new variable and opinion about the proposed wind farm was then tested. The correlation was significant at the 0.0005 level, although the coefficient is very low; nevertheless, it suggests that respondents who showed a greater level of attachment tended to oppose the wind farm (Table 5.54) more often than those who were not attached.

Table 5.54 Correlations between "opinion about the locally proposed wind farm" and "place_attachment2"

			Opinion about the locally proposed wind farm
Kendall's tau_b	Place_attachment2	Correlation coefficient	0.192
		Sig. (2-tailed)	0.000
		N	307

In order to account for any differences in correlations between opinions about the wind farm and perceptions of attachment to physical and social places, the individual correlations are reported (Tables 5.55–5.56).

Contextual factors

In this section, the correlations regarding the factors whose frequencies were presented earlier in this chapter are introduced.

Trust

When the variable "trust towards the developers" was tested with the opinion about the wind farm, a significant correlation was found at the 0.0005 level,

Table 5.55 Correlations between "opinion about the locally proposed wind farm" and "place attachment—physical place"

			Opinion about the locally proposed wind farm
Kendall's tau_b	Place attachment—physical place	Correlation coefficient	0.201
		Sig. (2-tailed)	0.000
		N	305

Table 5.56 Correlations between "opinion about the locally proposed wind farm" and "place attachment—social place"

			Opinion about the locally proposed wind farm
Kendall's tau_b	Place attachment—social place	Correlation coefficient	0.172
		Sig. (2-tailed)	0.001
		N	303

Table 5.57 Correlations between "opinion about the locally proposed wind farm" and "trust towards developers"

			Opinion about the locally proposed wind farm
Kendall's tau_b	*Trust towards developers*	Correlation coefficient	0.622
		Sig. (2-tailed)	0.000
		N	304

which shows that those who distrusted developers were also against the proposed wind farm (Table 5.57).

Information received

Equally significant at the 0.0005 level was the correlation between information provided about the wind farm project and opinion about the wind farm. Respondents who believed that they had not been fully informed about the proposed wind farm tended to oppose its construction (Table 5.58).

Commercial vs co-operative scheme

A weak significant (p<0.05) correlation was found between type of ownership, i.e. whether the wind farm was a commercial enterprise or a co-operative scheme, and opinions about the wind farm (Tables 5.59–5.61). It appears that residents who lived in an area where a co-operative scheme was proposed tended to be slightly more negative about the proposed wind farm. It should be emphasised, however, that while this relationship was of little significance, only a minority of respondents were actually aware of the scheme proposed in their area.

The co-operative model

This section presents the correlations regarding a series of items, whose frequencies have been earlier shown in this chapter and whose purpose was to

Table 5.58 Correlations between "opinion about the locally proposed wind farm" and "information about the wind farm"

			Opinion about the locally proposed wind farm
Kendall's tau_b	*Information about the wind farm*	Correlation coefficient	0.300
		Sig. (2-tailed)	0.000
		N	305

Table 5.59 Cross tabulation between the variables "opinion about the locally proposed wind farm" and "scheme of ownership co-op/commercial"

| | | | Scheme of ownership co-op/commercial | | Total |
			Co-op scheme	commercial scheme	
Opinion about locally proposed wind farm	Strongly disagree	Count	46	16	62
		% within scheme of ownership co-op/ commercial	23.7%	14.0%	20.1%
	Disagree	Count	21	10	31
		% within scheme of ownership co-op/ commercial	10.8%	8.8%	10.1%
	Neither agree nor disagree	Count	41	42	83
		% within scheme of ownership co-op/ commercial	21.1%	36.8%	26.9%
	Agree	Count	54	33	87
		% within scheme of ownership coop/ commercial	27.8%	28.9%	28.2%
	Strongly agree	Count	32	13	45
		% within scheme of ownership co-op/ commercial	16.5%	11.4%	14.6%
Total		Count	194	114	308
		% within scheme of ownership co-op/ commercial	100.0%	100.0%	100.0%

measure positive and negative attitudes towards a number of statements regarding co-operative schemes. A scale variable, named "co-opschemeopinion" was built, which ranked responses ranging from a score of 7 (only negative opinions about the co-operative scheme) to 35 (only positive opinions about the scheme).

The correlation between the new variable and opinion about the wind farm was then tested using the Kendall's tau test (2-tailed), because it is plausible to think that the direction of the relationship could be both ways. The result shows a significant correlation at the 0.0005 level with a positive coefficient

Table 5.60 Chi-square tests

	Value	Df	Asymp. sig. (2-sided)
Pearson chi-square	11.521[a]	4	0.021
Likelihood ratio	11.524	4	0.021
Linear-by-linear association	0.616	1	0.433
N of valid cases	308	–	–

Notes
a: no cells (0.0%) have an expected count below 5. The minimum expected count is 11.47.
Df: degrees of freedom.
Asymp. sig.: asymptotic significance.

Table 5.61 Symmetric measures

		Value	Approx. sig.
Nominal by nominal	Phi	0.193	0.021
	Cramer's V	0.193	0.021
N of valid cases		308	–

(Table 5.62), meaning that respondents who viewed the wind farm proposal negatively tended to evaluate the co-operative scheme negatively as well.

Investment in the co-operative scheme

The questionnaire asked respondents whether they would invest in a local wind farm co-operative at a price of £250 per share. As imagined, answers differed depending on whether respondents supported or opposed the project. In fact, a significant correlation at 0.0005 level was found, but the Cramer's V coefficient, which ranges from 0 to 1, was 0.53. This showed that those who opposed the project leaned towards not investing (Tables 5.63–5.65).

A number of statements followed in the questionnaire (questions 15.1–15.8, see Appendix A) that surveyed motives for the choice of investing. These scores will be analysed later in this chapter using multivariate correlation tests.

Table 5.62 Correlations between "opinion about the locally proposed wind farm" and "co-operative scheme opinion"

			Co-opschemeopinion
Kendall's tau_b	*Opinion about the locally proposed wind farm*	Correlation coefficient	0.555
		Sig. (2-tailed)	0.000
		N	282

Table 5.63 Cross tabulation between the variables "decision to invest" and "opinion about the locally proposed wind farm"

			Opinion about the locally proposed wind farm					Total
			Strongly disagree	Disagree	Neither agree nor disagree	Agree	Strongly agree	
Decision to invest	No	Count	52	21	46	26	5	150
		% within decision to invest	34.7%	14.0%	30.7%	17.3%	3.3%	100.0%
		% within opinion about locally proposed wind farm	92.9%	75.0%	61.3%	35.6%	12.8%	55.4%
	Yes	Count	4	7	29	47	34	121
		% within decision to invest	3.3%	5.8%	24.0%	38.8%	28.1%	100.0%
		% within opinion about locally proposed wind farm	7.1%	25.0%	38.7%	64.4%	87.2%	44.6%
Total		Count	56	28	75	73	39	271
		% within decision to invest	20.7%	10.3%	27.7%	26.9%	14.4%	100.0%

Table 5.64 Chi-square tests

	Value	Df	Asymp. sig. (2-sided)
Pearson chi-square	77.384[a]	4	0.000
Likelihood ratio	87.238	4	0.000
Linear-by-linear association	75.853	1	0.000
N of valid cases	271	–	–

Notes
a: no cells (0.0%) have an expected count below 5. The minimum expected count is 12.50.
Df: degrees of freedom.
Asymp. sig.: asymptotic significance.

Table 5.65 Symmetric measures

		Value	Approx. sig.
Nominal by nominal	Phi	0.534	0.000
	Cramer's V	0.534	0.000
N of valid cases		271	–

Opinion about the co-operative scheme

A question was posed about respondents' overall opinion of co-operative schemes (question 16, see Appendix A). This was likely to relate to opinion regarding the wind farm, so the correlation was tested with a 2-tailed test because it is not obvious which one of the two was the dependent variable (Table 5.66). The correlation is significant at 0.0005 level and the coefficient is 0.61. This means that those who disapproved of the wind farm tended to consider the co-operative model a bad idea too.

Preferred scheme: co-operative vs community fund

Respondents were asked about their preferred scheme of choice between a co-operative scheme, a community fund scheme or no preference.

Table 5.66 Correlations between "opinion about the locally proposed wind farm" and "co-operative idea opinion"

			Co-operative idea opinion
Kendall's tau_b	*Opinion about locally proposed wind farm*	Correlation coefficient	0.612
		Sig. (2-tailed)	0.000
		N	287

Table 5.67 Cross tabulation between the variables "scheme of choice" and "Opinionwf3items"

			Opinionwf3items			Total
			Disagree + strongly disagree	Neither agree nor disagree	Agree + strongly agree	
Scheme of choice	Community wind farm co-op	Count	19	21	65	105
		% within scheme of choice	18.1%	20.0%	61.9%	100.0%
		% within opinionwf3items	25.7%	26.6%	53.7%	38.3%
	Community fund	Count	23	21	27	71
		% within scheme of choice	32.4%	29.6%	38.0%	100.0%
		% within opinionwf3items	31.1%	26.6%	22.3%	25.9%
	No preference	Count	32	37	29	98
		% within scheme of choice	32.7%	37.8%	29.6%	100.0%
		% within opinionwf3items	43.2%	46.8%	24.0%	35.8%
Total		Count	74	79	121	274
		% within scheme of choice	27.0%	28.8%	44.2%	100.0%

Table 5.68 Chi-square tests

	Value	Df	Asymp. sig. (2-sided)
Pearson chi-square	23.369[a]	4	0.000
Likelihood ratio	23.578	4	0.000
Linear-by-linear association	16.451	1	0.000
N of valid cases	274	–	–

Notes
a: no cells (0.0%) have an expected count below 5. The minimum expected count is 19.18.
Df: degrees of freedom.
Asymp. sig.: asymptotic significance.

Table 5.69 Symmetric measures

		Value	Approx. sig.
Nominal by nominal	Phi	0.292	0.000
	Cramer's V	0.207	0.000
N of valid cases		274	–

Again, it made sense to test this question in relation to opinions about the wind farm. The resulting correlation was significant at the 0.01 level, showing that respondents opposing, or neither opposing nor supporting, the wind farm were more likely to express "no preference", while those that supported the wind farm were predominantly supportive of the co-operative scheme (Tables 5.67–5.69).

Summary table of correlations

Table 5.70 summarises the correlations presented earlier. It shows only significant correlations between the tested independent variables and the variable "opinion about the locally proposed wind farm". The variables are ordered by the size of the coefficient of correlation starting with the greater positive coefficient and ending with the greater negative coefficient.

Multivariate analysis

In the following sections, the analysis continues with some multivariate tests and, particularly, ordinal regression and logistic binomial regression. Before looking in detail at the tests, it is worth noting that a key assumption such as multicollinearity has been tested for each model, generating correlations matrixes, again using the Kendall's tau statistic (all the variables involved being ordinal). These did not result in any correlation which was considered of such magnitude (0.8 or more) (Field, 2009; ReStore, 2011a) to cause concern. Thus, the matrixes are not presented along with the model tests.

Table 5.70 Summary of significant correlations with the variable "opinion about the locally proposed wind farm"

Independent variable	Correlation coefficient (test performed)	Significance
"Benefitscostsvalue"[1]	0.731 (Kendall's tau)	0.0005
	0.855 (Spearman's rho)	0.0005
Local benefits/disadvantages	0.684 (Kendall's tau)	0.0005
Wind farm looks bad on landscape	0.654 (Kendall's tau)	0.0005
Trust toward developers	0.622 (Kendall's tau)	0.0005
Co-operative idea opinion	0.612 (Kendall's tau)	0.0005
Wind farm harms local health	0.571 (Kendall's tau)	0.0005
Wind farm improves local economy	0.567 (Kendall's tau)	0.0005
"Co-opschemeopinion"	0.555 (Kendall's tau)	0.0005
Wind farm is noisy	0.537 (Kendall's tau)	0.0005
Decision to invest	0.534 (Cramer's V)	0.0005
Wind farm helps fuel dependency	0.532 (Kendall's tau)	0.0005
Wind farm brings down property prices	0.526 (Kendall's tau)	0.0005
Wind farm attract tourists	0.505 (Kendall's tau)	0.0005
Wind farm helps climate change	0.501 (Kendall's tau)	0.0005
Wind farm generates costlier electricity	0.385 (Kendall's tau)	0.0005
Awareness of proposed wind farm	0.353 (Cramer's V)	0.0005
Information about wind farm	0.300 (Kendall's tau)	0.0005
Seeing wind farm site from home	0.272 (Cramer's V)	0.0005
"Awarescheme3"[2]	0.219 (Cramer's V)	0.001
Scheme of choice	0.207 (Cramer's V)	0.0005
Scheme of ownership	0.193 (Cramer's V)	0.021
"Place_attachment2"[3]	0.192 (Kendall's tau)	0.0005
Environment vs economy trade-off	0.157 (Kendall's tau)	0.002
Number of household members	0.114 (Kendall's tau)	0.013
Knowledge of wind energy "knowwind"	–0.078 (Kendall's tau)	0.051
Estimation of income per family member	–0.111 (Kendall's tau)	0.014
How often see wind farm site[4]	–0.123 (Kendall's tau)	0.047

Notes
1 As mentioned previously, this variable was obtained by creating a new variable from the scores of each respondent to the nine items in the questionnaire that concerned the perceived costs and benefits of the wind farm.
2 Again, the "opinionwf3items" variable was used and, in order to execute Cramer's V test correctly, the original variable of awareness about the proposed scheme was collapsed into a dichotomous variable.
3 As explained, "place_attachment2" is a purpose-built scale which includes both answers on the physical and social attachment items.
4 Tested vs the "opinionwf3items" variable, which had collapsed the 5-point Likert scale of the original variable into a 3-point scale.

Factors influencing acceptability of wind farms

From the summary table of correlations presented in the previous section (Table 5.70), it is possible to proceed to select some variables that will be used to perform an ordinal regression analysis.

Ordinal regression

"Ordinal regression", as presented by Norušis (2011), analyses data, i.e. predictors (independent variables) and the outcome variable (dependent variable), which are of ordinal nature. In fact, the researcher can perform a multivariate regression analysis retaining the ordinal information of the variable, i.e. the scores ordered, for example, on a Likert scale (as in this study), along a continuum that represents the concept measured. This removes the need to consider the variables improperly as interval variables in order to perform a linear regression (Bryman and Cramer, 2009; Field, 2009).

Selection of variables

With regard to the selection of variables to be included in the ordinal regression tests performed, these have been selected on the basis of the research questions; but, in this study, it was also necessary to pay attention at the number of predictors in relation to the number of cases collected. Field (2009) warns that the sample size must be high to detect even small effects, particularly when the number of predictors is large.

It makes sense, therefore, to limit the number of predictors to those that have shown a significant correlation and a correlation coefficient which is of at least "modest strength", with a value between 0.40 and 0.69, as defined by Cohen and Holliday (cited in Bryman and Cramer, 2009).

The first research question that I posed was: "Which factors influence acceptability of wind farms? How do they relate to each other?".

Clearly, to answer this question using the ordinal regression test earlier introduced, it was necessary to select the "opinion about the wind farm variable", (itself an ordinal variable), as the dependent variable of the test. Considering the size of their correlation coefficient, the following variables, bearing a significant correlation with the dependent variable, were selected to be included in the first ordinal regression test:

1   trust towards developers
2   local benefits/disadvantages
3   WF harms local health
4   WF helps climate change
5   WF looks bad on the landscape
6   WF improves the local economy
7   WF brings down property prices

8   WF is noisy
9   WF helps with fuel dependency.

Test statistics

In order to carry out an ordinal regression that limited as much as possible the chances of empty cells resulting from the combination of the variables' single modes of answer, which could eventually lead to reduced reliability of the "goodness of fit" statistics test of the model (Norušis, 2011), the variables were transformed from 5-point to 3-point scales.

Nevertheless, SPSS returned the following warning with the results of the test: "There are 364 (64,9%) cells (i.e., dependent variable levels by observed combinations of predictor variable values) with zero frequencies." This is not unusual when many variables are combined in the test. Usually, SPSS indicates if the number of empty cells is compromising specific outputs, and this was not the case here. Previously, I had tried to run the test with variables in the form of 5-point Likert scales, and the warning was: "The log-likelihood value is practically zero. There may be a complete separation in the data. The maximum likelihood estimates do not exist." This indicated that the test had instead collapsed.

In this case, it is possible to suppose that the high number of empty cells might be due to a polarisation of views regarding the wind farm, with fairly consistent answers across the sample between supporters and opponents of the project, i.e. supporters and opponents had consistently opposite views regarding statements about the advantages and disadvantages of the proposed wind farm.

To turn to the tables produced by the test, the first is the "-2 Log Likelihood test", which tells us whether our model is more accurate in explaining the outcome of the dependent variable, opinion about the proposed wind farm ("opinionwf3items"), than the base model using purely predictions based on the marginal probabilities of the outcome categories (ReStore, 2011b) (Table 5.71).

The chi-square statistic is significant at the 0.0005 level meaning that the model offers a better fit to the data than the base model.

The next statistic presented by SPSS is "goodness of fit", which relates to the correspondence between the data collected and the data predicted by the model (Table 5.72). The null hypothesis is that the fit is good, therefore it should not be

Table 5.71 Model fitting information

Model	−2 log likelihood	Chi-square	Df	Sig.
Intercept only	550.259	–	–	–
Final	198.840	351.419	18	0.000
Link function: logit				

Note
Df: degrees of freedom.

Table 5.72 Goodness of fit

	Chi-square	Df	Sig.
Pearson	259.687	354	1000
Deviance	183.448	354	1000
Link function: logit			

Note
Df: degrees of freedom.

rejected and the significance level of the corresponding chi-square statistic should be p>0.05. In this case, the null hypothesis would not be rejected, pointing to a supposedly good fit of the model.

ReStore (2011b) authors recommend relying on the "pseudo R^2 statistics" instead to assess the fitness of the model when many cells are empty. Pseudo R^2 shows the proportion of variance of the dependent variable explained by the model. In this case, as shown in Table 5.73, the Nagelkerke statistic, one of the most widely used (ReStore, 2011a), reports a value corresponding to 83 per cent of variance explained by the model.

The last test that is presented here before going into further detail about the model is the "test of parallel lines". This test aims to verify the necessary assumption of ordinal regression (also named "assumption of proportional odds") that states that each independent variable has the same effect across its different thresholds (ReStore, 2011b). This test is considered to be of an anti-conservative nature, i.e. often rejecting the assumption, which is the null hypothesis of the test, whenever there are many variables, the sample size is large or there is a continuous independent variable in the model (ReStore, 2011b).

In this case, the null hypothesis is not rejected because the significance level exceeds the 0.05 level; therefore, the null hypothesis is largely confirmed, i.e. the assumption of proportional odds is confirmed[1] (Table 5.74).

The following part of the statistical analysis goes into further detail, assessing the relative importance of each independent variable in the model. It is important to note that the dependent variable "opinion about the locally proposed wind farm", which is represented in the tables by the variable "opinionwf3items", is coded with "1" expressing disagreement, "2" for neither agreement nor disagreement and "3" for agreement. The independent variables are all coded from 1 to 3

Table 5.73 Pseudo R-square

Cox and Snell	0.734
Nagelkerke	0.831
McFadden	0.616
Link function: logit	

Table 5.74 Test of parallel lines[a]

Model	–2 log likelihood	Chi-square	Df	Sig.
Null hypothesis	198.840	–	–	–
General	170.664	28.177	18	0.059

The null hypothesis states that the location parameters (slope coefficients) are the same across response categories.

Notes
a: link function: logit.
Df: degrees of freedom.

starting either with disagreement or agreement but always with the answer "1" corresponding to a position most likely to be conducive to disagreement with the proposed wind farm. For example, the variable "Helpcc3" refers to the statement regarding the wind farm helping with climate change and the response number "1" is expressing disagreement.

As it is possible to see from Table 5.75, all the answers expressing a negative stance related to the wind farm had a negative significant relationship with agreement towards the wind farm. The variables "Wfeconom3" (the statement about the wind farm improving the local economy) and "Propprices3" (the statement about the devaluation of property prices due to the wind farm) do not have significant results in this test. However, the opposite is true for the Kendall's tau correlation tests presented earlier in the chapter, i.e. controlling for the variables included in the model, the variable "Wfeconom3" appears to lack statistical significance in its relationship with the dependent variable. From what can be seen from the data, there are four variables that show significant negative relationships with coefficients that are higher than the others. These are: trust towards the developers, local benefits/disadvantages, WF harms local health and WF helps climate change—all with negative coefficients, ranging from –2.5 (trust3) to –2.2 (helpcc3).

As it is noted in the literature (ReStore, 2011b) taking the exponent of the logit will return the odd ratio. So, in the case of respondents who disagree with the statement about trust towards the developers, $\exp(-2.52) = 0.080$, which means that they are less than one-tenth likely to agree with the wind farm compared to those who trust the developers (i.e. agree with the statement about having trust in the developers). Similar ratios are clearly returned from the other variables presenting coefficients of similar magnitude.

In Chapter 2, Diekmann and Preisendörfer's (2003) so-called "low-cost hypothesis" was presented, which argues that pro-environmental attitudes are mostly effective in shaping behaviours in low-cost situations.

This study's hypothesis was tested by performing correlation tests using the selected cases that answered "neither benefits nor disadvantages" when asked about their perception of local benefits or disadvantages. This assumed that those who had otherwise leaned towards disadvantages or advantages would have perceived the situation as "high-cost" or even "high-benefit".

Table 5.75 Parameter estimates

		Estimate	S. E.	Wald	Df	Sig.	95% confidence interval	
							Lower bound	Upper bound
Threshold	[Opinionwf3items = 1.00]	-8.890	1.036	73.679	1	0.000	-0.920	-6.860
	[Opinionwf3items = 2.00]	-4.570	0.763	35.842	1	0.000	-6.067	-3.074
Location	[Trust 3=1.00]	-2.525	0.732	11.899	1	0.001	-3.960	-1.090
	[Trust 3=2.00]	-1.155	0.650	3.161	1	0.075	-2.428	0.118
	[Benefdisadv_local 3 = 1.00]	-2.350	0.619	14.409	1	0.000	-3.564	-1.137
	[Benefdisadv_local 3 = 2.00]	-0.413	0.500	0.682	1	0.409	-1.394	0.568
	[Wfharms 3 = 1.00]	-2.398	1.126	4.532	1	0.033	-4.606	-0.190
	[Wfharms 3 = 2.00]	-1.073	0.458	5.495	1	0.019	-1.970	-0.176
	[Helpcc 3 = 1.00]	-2.198	0.579	14.394	1	0.000	-3.333	-1.062
	[Helpcc 3 = 2.00]	-1.288	0.482	7.158	1	0.007	-2.232	-0.345
	[Lookbad 3 = 1.00]	-1.423	0.561	6.431	1	0.011	-2.524	-0.323
	[Lookbad 3 = 2.00]	-0.70	0.491	0.020	1	0.887	-1.032	0.892
	[Wfeconom 3 =1.00]	-0.353	0.584	0.365	1	0.546	-1.498	0.792
	[Wfeconom 3 = 2.00]	-0.930	0.507	3.358	1	0.067	-1.924	0.065
	[Propprices 3 = 1.00]	-1.032	0.578	3.189	1	0.074	-2.164	0.101
	[Propprices 3 = 2.00]	-0.832	0.517	2.596	1	0.107	-1.844	0.180
	[Wfnoise 3 = 1.00]	-1.463	0.652	5.025	1	0.025	-2.741	-0.184
	[Wfnoise 3 = 2.00]	-0.589	0.413	2.038	1	0.153	-1.398	0.220
	[Fueldepend 3 = 1.00]	-1.196	0.562	4.533	1	0.033	-2.296	-0.095
	[Fueldepend 3 = 2.00]	-0.473	0.476	0.986	1	0.321	-1.406	0.461

Notes
S. E.: standard error.
Wald: Wald chi-square test.
Df: degrees of freedom.
Sig.: significance of the test.

These cases were tested in correlations between the variable "opinion about the wind farm" and environmental attitudes variables, i.e. the variables "environment vs economy trade-off", "physical place attachment", "social place attachment" and "environmental citizenship level" which, with the exception of "physical place attachment", returned non-significant correlations. In the latter case the correlation was statistically significant at the 0.05 level and modest in its strength (0.2).

Later, after resetting SPSS to include all cases (respondents) we tried to include in four separate ordinal regression models all the four attitudinal variables. These did not present a significant relationship with the dependent variable "opinion about the locally proposed wind farm", nor did they increase the variance explained by the model or lead to a collapse of the model.

It should be noted, however, that the questions used to try to tap into environmental attitudes might have been too limited in their scope, and that already tested multi-item scales might have been more suitable instruments given the possibility of producing a lengthier questionnaire.

Factors influencing opinions about the co-operative model

The second research question that this research posed was: "Which factors influence participation in wind farm co-operatives? How do they relate to each other?". The questionnaire tried to capture the sample's views concerning a number of contentious statements about the co-operative model (questions 15.1–15.7 in the questionnaire, see Appendix A).

In order to assess how these issues could influence opinions about the wind farm co-operative model, and hence its acceptance, an ordinal regression test was conducted, which had as independent variables the items corresponding to questions 15.1–15.7 and, as a dependent variable, "opinion about community wind farm co-operative scheme", corresponding to question 16 in the questionnaire.

Even in this test, statistics relating to the fitness of the model presented results pointing to a good fit. The –2 Log Likelihood test presented a chi-square value significant at the 0.0005 level, hence showing a fit of the model which is better than for the basic model (Table 5.76).

Table 5.76 Model fitting information

Model	–2 log likelihood	Chi-square	Df	Sig.
Intercept only	734.791	–	–	–
Final	441.896	292.895	32	0.000
Link function: logit				

Note
Df: degrees of freedom.

The "goodness of fit" statistics presented a chi-square with a non-significant level of 1, meaning that the null hypothesis cannot be rejected and thus the model presents a good level of fit.

The pseudo R^2 test presents a fairly high level of variance explained by the model that for the Nagelkerke statistic is 70 per cent (Table 5.77).

The test of parallel lines (Table 5.78) results in a non-significant chi-square value, which means that the null hypothesis is not rejected, hence following the necessary assumption of proportional odds.[2]

Table 5.79 presents the coefficient estimates. As can be seen, four variables bear a significant relationship with the dependent variable—"co-op scheme is a ploy", "co-op revenue will benefit community", "co-op will offer the worst compensation" and "co-op will create social capital".

Not surprisingly, respondents who strongly agreed with the statement, "The co-op scheme is a ploy" were less than one hundredth times less likely to consider the co-op a good idea than those who thought it was not a ploy, ranging from "the co-op is a very bad idea" to "the co-op is a very good idea".

Similarly, those who strongly disagreed with the statement, "Co-op revenue will benefit the community" were about 1/50 times likely to move towards more support regarding the co-operative idea.

Those who strongly agreed that the co-op offered the worst compensation were less than 1/10 likely to approve the co-operative scheme. Similarly, those who strongly disagreed with the statement that the co-operative would build social capital were about 1/50 likely to approve the wind farm co-operative scheme compared with those who believed the opposite.

Table 5.77 Pseudo R-square

Cox and Snell	0.665
Nagelkerke	0.704
McFadden	0.378
Link function: logit	

Table 5.78 Test of parallel lines[a]

Model	−2 log likelihood	Chi-square	Df	Sig.
Null hypothesis	441.896	−	−	−
General	355.742	86.153	96	0.754

The null hypothesis states that the location parameters (slope coefficients) are the same across response categories.

Notes
a: link function: logit.
Df: degrees of freedom.

Table 5.79 Parameter estimates

	Estimate	S.E.	Wald	Df	Sig.	95% confidence interval	
						Lower bound	Upper bound
Threshold							
[co-op_idea = 1.00]	-11.281	2.402	22.053	1	0.000	-15.990	-6.573
[co-op_idea = 2.00]	-9.310	2.372	15.400	1	0.000	-13.960	-4.660
[co-op_idea = 3.00]	-6.720	2.352	8.163	1	0.004	-11.329	-2.110
[co-op_idea = 4.00]	-2.691	2.327	1.337	1	0.247	-7.251	1.869
Location							
[c_ploy=1.00]	-4.705	1.042	20.401	1	0.000	-6.746	-2.663
[c_ploy=2.00]	-3.250	0.955	11.582	1	0.001	-5.121	-1.378
[c_ploy=3.00]	-2.367	0.931	6.466	1	0.011	-4.191	-0.543
[c_ploy=4.00]	-1.123	0.894	1.578	1	0.209	-2.876	0.629
[c_revenue=1.00]	-3.867	1.579	6.000	1	0.014	-6.962	-0.773
[c_revenue=2.00]	-4.566	1.472	9.629	1	0.002	-7.451	-1.682
[c_revenue=3.00]	-4.228	1.426	8.788	1	0.003	-7.023	-1.433
[c_revenue=4.00]	-2.797	1.378	4.121	1	0.042	-5.497	-0.097
[c_divide=1.00]	2.046	1.484	1.903	1	0.168	-0.861	4.954
[c_divide=2.00]	2.154	1.391	2.396	1	0.122	-0.573	4.880
[c_divide=3.00]	2.306	1.391	2.749	1	0.097	-0.420	5.032
[c_divide=4.00]	2.497	1.426	3.065	1	0.080	-0.299	5.292
[c_undecided=1.00]	0.213	1.715	0.015	1	0.901	-3.149	3.575
[c_undecided=2.00]	0.721	1.563	0.212	1	0.645	-2.344	3.785
[c_undecided=3.00]	0.375	1.549	0.059	1	0.808	-2.660	3.411

	Estimate	S.E.	Wald	Df	Sig.		
[c_undecided=4.00]	0.228	1.543	0.022	1	0.883	-2.797	3.253
[c_compens=1.00]	-2.637	1.147	5.289	1	0.021	-4.885	-0.390
[c_compens=2.00]	-1.548	1.052	2.164	1	0.141	-3.610	0.514
[c_compens=3.00]	-1.354	1.051	1.660	1	0.198	-3.415	0.706
[c_compens=4.00]	-1.909	1.098	3.022	1	0.082	-4.061	0.243
[c_socap=1.00]	-3.292	1.093	9.070	1	0.003	-5.434	-1.149
[c_socap=2.00]	-3.010	0.922	10.669	1	0.001	-4.817	-1.204
[c_socap=3.00]	-1.864	0.867	4.627	1	0.031	-3.563	-0.166
[c_socap=4.00]	-1.274	0.822	2.402	1	0.121	-2.886	0.337
[c_antis=1.00]	0.024	1.826	0.000	1	0.990	-3.556	3.604
[c_antis=2.00]	-0.206	1.783	0.013	1	0.908	-3.701	3.288
[c_antis=3.00]	0.817	1.796	0.207	1	0.649	-2.703	4.337
[c_antis=4.00]	0.298	1.773	0.028	1	0.866	-3.177	3.773
[c_nodiff=1.00]	-1.104	1.151	0.920	1	0.337	-3.359	1.152
[c_nodiff=2.00]	-0.011	1.096	0.000	1	0.992	-2.159	2.137
[c_nodiff=3.00]	-0.555	1.091	0.259	1	0.611	-2.694	1.584
[c_nodiff=4.00]	0.374	1.155	0.105	1	0.746	-1.890	2.637

Notes
S. E.: standard error.
Wald: Wald chi-square test.
Df: degrees of freedom.
Sig.: significance of the test.

These results confirm that issues of trust (the co-operative is a ploy) and the perception of lack of sufficient financial/material benefits (poor compensation, little to no revenue for the community) influence opinions about a scheme that is seen by some researchers (Lipp and McMurtry, 2015) as improving the acceptability of wind farm developments.

Factors influencing the decision to invest in a co-operative wind farm

The following test aims to assess which factors most influence the decision to invest in a co-operative wind farm through a binomial logistic regression.

Binomial logistic regression

Binomial logistic regression is a necessary choice for a multivariate analysis when the dependent variable has a binary outcome, as in this case where the sample was asked whether or not they would invest (ReStore, 2011a). The test involved only 229 cases because in a large number of the returned questionnaires this section was not completed. Nevertheless, the sample size of 229 was sufficient to detect a medium effect (Field, 2009). The dependent variable, which asked about willingness to invest (question 14 in the questionnaire, see Appendix A), was coded as follows (Table 5.80):

The model tested, including the variables corresponding to questions 15.1–15.7, made a comparison with the baseline model that predicted the outcome purely on the basis of which category occurred most often in the data set, thereby achieving 53.7 per cent correct predictions.

All the variables introduced in the model were tested in one block, following Field (2009) who recommends this procedure unless sound motives supported by previous research would suggest otherwise.

The "omnibus test of model coefficients" tells if the new model has the –2 Log likelihoods statistic significantly reduced compared to the baseline model, which would mean that the new model is capable of explaining more of the variance in the outcome. As can be seen (Table 5.81), in this case the chi-square value is significant at the 0.0005 level.

The model summary (Table 5.82) provides the pseudo-R^2 statistic that presents a Nagelkerke value of 0.888, suggesting that the model explains 88 per cent of the variance in the outcome.

Table 5.80 Dependent variable encoding

Original value	Internal value
No	0
Yes	1

Table 5.81 Omnibus tests of model coefficients

		Chi-square	Df	Sig.
Step 1	Step	250.331	7	0.000
	Block	250.331	7	0.000
	Model	250.331	7	0.000

Note

Df: degrees of freedom.

Table 5.82 Model summary

Step	−2 log likelihood	Cox and Snell R-square	Nagelkerke R-square
1	65.867[a]	0.665	0.888

Note

a: estimation terminated at iteration number 8 because parameter estimates changed by less than 0.001.

The Hosmer and Lameshow test tells whether the model is a "good enough fit" (ReStore, 2011a) for the data, which corresponds to the null hypothesis; therefore, this was not rejected and p must be higher than 0.05, which was the case in this test (Table 5.83).

The classification table (Table 5.84) indicates the percentage of correctly predicted outcomes by the model that can be compared with the same statistic from the baseline model, which stands at about 54 per cent. The percentage of correct prediction of the model reached nearly 94 per cent, therefore presenting a dramatic improvement.

Looking in detail at the variables in the model (Table 5.85), it is noticeable that all but two have a significant relationship with the dependent variable. These correspond to the statements "I don't care about the wind farm and so I would not care about the co-operative" and "If people around me, in my community, would support it, so would I". The first statement presented a situation of utter disengagement, while the second attempted to capture the relevance of social pressure to participate in the project.

The two variables with a significant relationship that stand out in this analysis correspond to the statement, "I think that it would be a good investment opportunity"

Table 5.83 Hosmer and Lameshow test

Step	Chi-square	Df	Sig.
1	3.778	8	0.877

Note

Df: degrees of freedom.

Table 5.84 Classification table[a]

Observed			Predicted		
			Decision to invest		Percentage correct %
			No	Yes	
Step 1	Decision to invest	No	115	8	93.5
		Yes	6	100	94.3
	Overall percentage		–	–	93.9

Note
a: the cut value is 0.500.

(i_ok in Table 5.85), which presents a positive coefficient B of 2.3 that tells us that respondents who answered "strongly agree" are about ten times more likely to invest than those who strongly disagreed.

Similarly, those who strongly agreed with the statement "I couldn't afford to buy the shares" (named i_noafford), were likely to invest once out of ten times compared with those who strongly disagreed.

Interestingly, economic motives appeared to be the strongest, although immediately followed by opposition to the wind farm (i_oppose) and the wish to do something about climate change (i_envattitd). The least important was the variable that tapped into the sense of opportunity to shape the project ("I would be able to have a say in the development of the wind farm and its management", i_havesay), which showed that those who strongly agreed that they would have a role in the project were three times more likely to invest than those who strongly disagreed.

Table 5.85 Variables in the equation

		B	S. E.	Wald	Df	Sig.	Exp(B)
Step 1[a]	i_ok	2.328	0.530	19.290	1	0.000	10.257
	i_oppose	1.752	0.526	11.086	1	0.001	5.767
	i_envattitd	1.532	0.478	10.277	1	0.001	4.629
	i_noafford	−2.048	0.455	20.299	1	0.000	0.129
	i_socpressu	0.056	0.430	0.017	1	0.896	1.058
	i_nocare	0.022	0.505	0.002	1	0.965	1.023
	i_havesay	1.117	0.429	6.772	1	0.009	3.054
	Constant	−17.226	4.162	17.130	1	0.000	0.000

Notes
a: variable(s) entered on step 1: i_ok, i_oppose, i_envattitd, i_noafford, i_socpressu, i_nocare,i_havesay.
B: coefficient.
S. E.: standard error.
Wald: Wald chi-square test.
Df: degrees of freedom.
Sig.: significance of the test.
Exp(B): exponentiation of the B coefficient.

Final remarks

This chapter has presented the results of the postal survey of residents living within the vicinity of four proposed wind farm sites in Scotland and discussed their relative significance, as evaluated by appropriate statistical tests. The extent to which the results obtained support the theoretical framework developed in Chapter 2 is discussed in the next chapter.

Notes

1 The test initially came with a warning about its validity due to reaching the maximum number of iterations; it was therefore rerun with an increased number of iterations in the SPSS options tab. I couldn't find any advice against this, not even in the SPSS online guidelines, which simply recommend using a number of iterations of non-negative integer (IBM Knowledge Center, 2012). Marôco's (2007) simple advice is to increase the numbers of step-halving when a similar warning is shown about the maximum number of step-halving being reached.

2 Even in this case, as in the previous ordinal regression analysis, the first execution of this statistic returned a warning about its validity due to reaching the maximum number of step-halving. So this number was increased in the option tab of the SPSS ordinal regression test, as suggested by Marôco (2007).

References

Bryman, A. and Cramer, D. 2009. *Quantitative data analysis with SPSS 14, 15 & 16: A guide for social scientists*. London: Routledge.

Cohen, L. and Holliday, M. 1982. *Statistics for social scientists*, London: Harper & Row.

Diekmann, A. and Preisendörfer, P. 2003. Green and greenback: The behavioral effects of environmental attitudes in low-cost and high-cost situations. *Rationality and Society*, 15, 441–472.

Field, A. 2009. *Discovering statistics using SPSS*. London: Sage.

IBM Knowledge Center. 2012. *Ordinal regression options* [Online]. Available at: www.ibm.com/support/knowledgecenter/SSLVMB_21.0.0/com.ibm.spss.statistics.help/idh_plum_opt.htm [Accessed 10 March 2016].

Jones, C. R. and Eiser, J. R. 2009. Identifying predictors of attitudes towards local onshore wind development with reference to an English case study. *Energy Policy*, 37, 4604–4614.

Lipp, J. and McMurtry, J. J. 2015. *Benefits of renewable energy co-operatives: Summary of literature review from the Measuring the Co-operative Difference Research Network*. Measuring the Co-operative Difference Research Network, Canada. Available at: http://ec.msvu.ca:8080/xmlui/handle/10587/1608.

Marôco, J. 2007. Regressão ordinal. *Análise estatística com o SPSS Statistics* (3rd edn.). Lisbon: Edições Sílabo.

Nelsen, R. B. 2011. Kendall tau metric [Online]. *Encyclopedia of mathematics*. Available at: www.encyclopediaofmath.org/index.php?title=Kendall_tau_metric&oldid=12869 [Accessed 13 November 2014].

Norušis, M. 2011. Ordinal regression. *IBM SPSS statistics 19 advanced statistical procedures companion.* Boston, MA: Addison-Wesley.

ONS. 2001. *Census 2001.* Available at: www.statistics.gov.uk/census [Accessed 21 November 2006].

ReStore. 2011a. *Module 4: Multiple logistic regression.* National Centre for Research Methods, Southampton, UK. Available at: www.restore.ac.uk/srme/www/fac/soc/wie/research-new/srme/modules/mod4/module_4_-_logistic_regression.pdf [Accessed 20 February 2016].

ReStore. 2011b. *Module 5: Ordinal regression.* National Centre for Research Methods, Southampton, UK. Available at: www.restore.ac.uk/srme/www/fac/soc/wie/research-new/srme/modules/mod5/module_5_-_ordinal_regression.pdf [Accessed 21 February 2016].

The Electoral Commission. 2005. *Social exclusion and political engagement: Research report, November 2005.* The Electoral Commission, London. Available at: www.electoralcommission.org.uk/sites/default/files/pdf_file/Social-exclusion-and-political-engagement.pdf.

van der Horst, D. and Toke, D. 2010. Exploring the landscape of wind farm developments: Local area characteristics and planning process outcomes in rural England. *Land Use Policy,* 27, 214–221.

6 A theory of social acceptability of wind farms

Finding a place for the co-operative model

Ambitious international and European (European Parliament, 2018) renewable energy targets are pushing Europe (and the world) towards further exploitation of forms of renewable energy, of which wind is arguably one of the most widely available in many European countries, including the UK (DECC, 2011).

Research focusing on specific facets of wind energy schemes that could potentially increase the acceptability of proposed sites among local residents and provide broad social and economic benefits is particularly relevant, as the considerable amount of published papers on the topic demonstrate (e.g. Bell *et al.*, 2013; Batel and Devine-Wright, 2015; van Veelen and Haggett, 2016). In particular, community projects are considered to be a positive type of wind development; nevertheless, they are not devoid of a number of complications which make them less likely to succeed than purely commercial projects. There are a number of reasons for this that chiefly pertain to financial and social aspects (Haggett and Aitken, 2015).

Co-operatives are seen as one of the means to increasing acceptability (Lipp and McMurtry, 2015; Bauwens *et al.*, 2016,) along with community benefits schemes (Walker *et al.*, 2014). The chief advantages of co-operatives are the opportunities they provide for citizens to participate in ownership of the wind farm and hence in benefiting from its revenue, creating a social network and developing community skills. However, co-operatives are not without their disadvantages, such as the potential to create divisions between those who join them and those who do not (Haggett and Aitken, 2015).

This study refines theories on the social acceptability of renewable energy, focusing, in particular, on the role of the co-operative model in facilitating the social acceptability of wind farms by examining the perceptions of such schemes by local residents who would be affected by their development.

Can an integrated rational choice and attitudes framework explain acceptability of wind farms?

The idea that community benefits and co-operative schemes can increase acceptability recalls what has been hypothesised earlier in this book, in Chapters 1 and 2, about the importance of perceived costs and benefits in shaping pro-environmental behaviours and, in this specific case, the acceptability of locally proposed wind farms. The first research question asked: "Which factors influence acceptability of wind farms? How do they relate to one another?".

In an attempt to answer this question, Chapter 2 proposed an integrated rational choice and attitudes framework to explain how attitudinal factors, personal resources and contextual factors influence the acceptability of wind projects and encourage participation in a wind farm co-operative.

The low-cost hypothesis

In 2007, in the context of another research project, this same framework, using as a key underlying assumption the low-cost hypothesis of Diekmann and Preisendörfer (2003), was outlined in a conference paper (Pellegrini-Masini, 2007) and discussed with regard to household energy-saving behaviours. It was argued that individuals act, in respect of such behaviours, under the basic drive of unsatisfied needs, which vary between individuals in relation to their level of needs satisfaction. This coheres with well-established theories of needs (Maslow, 1987; Taormina and Gao, 2013). Individuals, then, choose household energy-saving behaviours by evaluating the perceived costs and benefits of their energy actions, which includes the benefit of cognitive consistency with pro-environmental attitudes that may be held.

More recently, several other authors have further elaborated on the "low-cost hypothesis" input on research about motives for pro-environmental behaviours, and have recognised the importance of low-cost vs high-cost situations (De Groot and Steg, 2009; Steg *et al.*, 2014) and the perception of cost and benefits in shaping acceptability of energy alternatives (Perlaviciute and Steg, 2014).

So, what do the empirical studies discussed in this book add to these efforts to frame pro-environmental behaviours theoretically?

Perceived costs and benefits

The postal survey of residents in the vicinity of four proposed sites in Scotland appeared to confirm the findings of the qualitative study of Westmill Wind Farm: the perception of costs and benefits associated with a proposed wind farm influences the decision whether to support the project.

This can be seen by the relatively high correlation coefficients of the variables "benefitscostsvalue" and "local benefits/disadvantages" with the variable "opinion about the proposed wind farm" (Table 5.70, p. 184), which resulted in the two highest bivariate correlations with the opinion about the wind farm variable. The summary table of bivariate correlations presented in Chapter 5 (Table 5.70, p. 184) shows that the second highest correlation coefficient is 0.684 ($p<0.000$) of the variable "local benefits/disadvantages" just after 0.731 ($p<0.000$) of the variable "benefitscostsvalue", a scale built on the items that asked respondents to score their agreement with specific hypothetical costs and benefits (of both local and non-local impact) related to the proposed wind farm.

Further, the ordinal regression analysis (Chapter 5) that had "opinion about the proposed wind farm" as its dependent variable revealed that, along with three other variables with similar coefficients ("trust towards developers", "the wind farm harms local health" and "the wind farm helps with climate change"), the perception of local costs and benefits was one of the main indicators of support or opposition. Thus, a subjective assessment of local costs and benefits influenced respondents' opinions about a proposed development. The higher respondents believed the costs of having a wind farm in their locale would be, the more they opposed it.

The ordinal regression analysis (Chapter 5) also revealed that "the wind farm will harm the health of my community" and "the wind farm will help to fight climate change" are two of the four major variables that influenced the acceptability of proposed wind farms. The first is clearly a major cost which, if perceived, discourages support, while the second is a wider benefit that will impact the global rather than the local environment. The latter benefit is thus different from the other three variables ("trust towards developers", "perception of local costs and benefits" and "the wind farm will harm the health of my community") and stands out in the ordinal regression test which pertains only to the local community level. Despite that, climate change is a central issue in the debate around the importance of the wider deployment of wind energy, and is arguably perceived by supporters as its major environmental benefit, while opponents see it as the main claim to be undermined.

Somewhat surprisingly, the ordinal regression analysis indicated that visual impact was relatively less important, albeit statistically significant, in influencing acceptability of proposed developments. It had a similar coefficient as the variables concerning the belief that the wind farm would be noisy and its contribution towards reducing national fuel dependency.

Other variables included in the test, which included the possible decrease in property values due to the presence of the wind farm and its possible positive contribution to the local economy, did not result in significant correlations with the dependent variable in the ordinal regression analysis, despite showing significant bivariate non-parametric correlations with the opinion about the wind farm variable.

Attitudinal variables

The questionnaire directly tested pro-environmental attitudinal variables by presenting respondents with a hypothetical trade-off between the economy and the environment, and using concepts such as place attachment (in its physical and social dimensions) and environmental citizenship (the object of research question number four).

Attitudinal variables did not fit in the ordinal model, even though several attempts were made, including separating the single variables "environmental citizenship level", "economy vs environment trade-off", "physical place attachment" and "social place attachment", which either resulted in the collapse of the model (in the case of place attachment variables) or in non-significant coefficients.

The bivariate correlations between these variables and opinion about the wind farm were significant (p<0.001) but with modest strength (x≤2) only for "economy vs environment trade-off", "physical place attachment" and "social place attachment", as reported in Chapter 5.

In addition, when correlation tests were carried out between these attitudinal variables and opinion about the wind farm, after selecting for the cases of those respondents who thought that the presence of the wind farm would bring neither local advantages nor disadvantages locally, the results showed non-significant correlations in all but one case. The only significant correlation (Kendall's tau test, p<0.05) was for the variable regarding physical place attachment (question 20, see Appendix A), which presented a modest coefficient of 0.204. At first glance, this could be interpreted as showing the marginal role of pro-environmental attitudes, environmental citizenship and place attachment in determining support for or opposition towards a locally proposed wind farm, even when respondents consider having a wind farm built in their locale to be "low-cost" situation.

Arguably, this observation could undermine Diekmann and Preisendörfer's (2003) assumption in the low-cost hypothesis that attitudes play a role in low-cost situations. However, in the specific case of acceptability of wind farms, as discussed in Chapter 1 there is the possibility that pro-environmental attitudes might play a role both in favour of and against a proposed wind farm; this depends on whether respondents are influenced by a type of local conservationist environmentalism or global environmentalism, in what has been named a "green on green" controversy (Warren *et al.*, 2005). Support for this interpretation comes from the evidence that for the same selected cases (i.e. those respondents who envisioned neither local advantages nor disadvantages), a significant positive correlation (Kendall's tau test, x = 0.389, p<0.000) was instead found between the variable regarding the statement "the wind farm will help to fight climate change" and opinion about the wind farm. Further, for the same respondents, a correlation was also found between the variable "looks bad on the landscape" and opinion about the wind farm (Kendall's tau test, x = 0.579, p<0.000). These were the largest correlations, for these selected respondents, within the set of correlation tests between each of the variables (questions 11–11.8, see Appendix A) measuring specific costs and benefits of the proposed wind farm and the opinion about the wind farm variable.

Both of these correlations concern the wind farm's different environmental impacts: one on the global environment, specifically on climate change, and the other on the local environment, in particular the landscape. Thus, it seems plausible that environmental attitudes are influential in determining the support or opposition of respondents who consider the wind farm project a low-cost situation, but without univocally directing them towards support or opposition. This hypothesis is further supported by the frequencies of answers on the variable regarding environmental citizenship, (questions 18–18.3, Appendix A). Among respondents who chose the most pro-environmental option in the variable proposing a trade-off between the economy and the environment (questions

19–19.3, Appendix A), a considerable number (27 per cent) actually appeared more sensitive to protecting the local environment over the global environment (61.5 per cent), thereby showing that a sizable proportion of self-represented environmentalists are more concerned with defending the integrity of their local environment than that of the global environment.

Contextual factors

Previous literature has suggested that conditions specific to the social context, such as issues related to procedural justice and community schemes (including the co-operative model), are among the contextual factors that influence the acceptability of wind energy. These are discussed in the following paragraphs.

Trust

The ordinal regression analysis revealed that "trust in the developers" is among the group of variables that most influence support for or opposition to proposed wind farms. Recent research has confirmed the importance of this issue (Haggett and Aitken, 2015; Kalkbrenner and Roosen, 2016). Trust can be considered a contextual social resource (Misztal, 2013) which, depending on whether relations with the developer are positive and trust-building or negative and lead to a trust deficit, magnifies or reduces the perceived costs and benefits. In fact, a correlation was found in this study between trust and the perception of local advantages and disadvantages associated with the proposed local wind farm with a 0.000 significance level and a coefficient of 0.6 (Kendall's tau correlation test), therefore illustrating that respondents who were most mistrustful of the developers tended to see the fewest possible local benefits deriving from the proposed wind farm. The importance of trust was also confirmed by the ordinal regression analysis, mentioned above, which, along with three other significant variables, presented the highest coefficient of correlation with opinion about the proposed wind farm. In the bivariate correlation test, again, trust was among the four variables showing the highest regression coefficient with the opinion about the wind farm variable (specifically 0.622, significant at 0.000 level, see Table 5.70, p. 184).

The issue of trust also appeared as particularly important in the case of Westmill Wind Farm, as seen in the analysis of the qualitative study (see Chapter 4), thus suggesting that even a co-operative community-led scheme might raise issues of trust.

This was, in fact, confirmed by the postal survey. Question 13 (see Appendix A) asked respondents about their level of agreement with the statement that the co-operative was just a ploy to buy residents' consensus. A relative majority of respondents (44 per cent) "agreed" or "strongly agreed" (see p. 146, Chapter 5). Furthermore, when testing for correlation between question 13 and opinion about the co-operative scheme, a significant correlation (Kendall's tau test, $p < 0.000$, x = 0.613) was found, indicating that respondents who considered the co-operative scheme "a ploy" to buy consensus were more inclined to consider it a bad idea. Finally, the ordinal regression analysis (see pp. 190–194), which had opinion about the co-operative scheme as its dependent variable (question 16, Appendix A), also showed that the belief that the co-operative was a ploy had the highest significant coefficient of all the variables included. Once again, this confirms that the issue of trust, even with regard to the co-operative scheme, was the most important in shaping opinions about the scheme.

Information

In Chapter 2, the provision of information about the wind farm was indicated as an element of procedural justice in relation to wind farm siting.

In the study presented in Chapter 5, information (question 9, Appendix A) appeared to be correlated with opinion about the proposed wind farm, implying that respondents who did not believe that they had been fully informed about the proposed development tended to have a more negative opinion about it.

Unsurprisingly, a correlation was also found between the information variable and the variable concerning local benefits and disadvantages (Kendall's tau test, $x = 0.275$, $p < 0.000$), suggesting that the feeling of having been poorly informed negatively influenced respondents' appraisal of the wind farm's future benefits and disadvantages.

The co-operative model

Earlier chapters included the co-operative model along with community benefits schemes within the group of contextual factors.

With regard to the co-operative model, research question 2 (Chapter 3) asked: "Which factors influence participation in wind farm co-operatives? How do they relate to one another?". Research question 3 was: "Is the co-operative scheme effective in eliciting local communities' participation and in overcoming opposition towards wind developments? Why?".

Looking in more detail at the findings regarding the co-operative model, and particularly the second ordinal regression analysis presented in Chapter 5, it is apparent that "trust" is the key issue affecting the dependent variable, opinion about the co-operative scheme, as already discussed.

To a lesser extent, the ordinal regression indicated that the belief that the co-operative would give the community the opportunity to benefit from the wind farm revenue plays a role in shaping the judgement of the scheme. This is not surprising, and is actually one of the co-operative model's strengths, a belief held by 51 per cent of respondents and with which only around 18 per cent disagreed.

Following this variable, with smaller coefficients, there are the variables regarding the possibility of the scheme forming a stable community network ("social capital") and the belief that the co-operative would offer the worst compensation to those who oppose it.

Considerable numbers of respondents agreed with both of these statements, outnumbering those who disagreed with them. It is interesting to note that about 41 per cent of respondents agreed that the co-operative scheme would create a stable network inside the community—one of the scheme's major benefits after revenue. Also, another important finding is recognition of the shortcomings of the model, i.e. the fact that it might not be inclusive, because it offered poor compensation to those that oppose it. This finding is not unexpected: about 37 per cent of respondents agreed with this statement, again outnumbering those who disagreed.

The ordinal regression test revealed that the suitability of the scheme for winning the support of undecided residents (question 13.3, Appendix A) was not a statistically significant factor in shaping opinion about the co-operative scheme. A very modest significant correlation, (Kendall's tau test, $x = 0.10$, $p < 0.05$), was found between the two variables when tested. The vast majority of respondents (about 44 per cent) neither agreed nor disagreed with the statement, while about 30 per cent disagreed, outnumbering those who agreed by a few per-centage points. Very interestingly, in response to a separate question suggesting that the co-operative scheme would not make any difference in terms of support for the wind farm, an overwhelming majority (65.5 per cent) of respondents replied that they agreed, while just 9.5 per cent disagreed. Nevertheless, this variable was not found to be in a statistically significant correlation with the opinion about the co-operative scheme variable in either the Kendall's tau test or the ordinal regression analysis, meaning that this belief did not appear to affect respondents' judgement about the scheme itself.

About 42.5 per cent of respondents agreed that the co-operative would have created a permanent divide within the community between those who joined it and those who opposed the wind farm, while about 20 per cent disagreed. This variable had a highly significant ($p<0.001$) bivariate correlation with the variable "opinion about the co-operative", although it was of moderate strength (0.47), while in the ordinal regression test it did not reach the level of significance ($p < 0.05$).

When respondents were asked whether they would invest in the co-operative scheme (question 14, Appendix A), 56 per cent answered "No" and 44 per cent "Yes". As reported in Chapter 5, a binomial logistic regression revealed that the main reasons for this decision were of an economic nature: either the belief that the co-operative scheme would be a good investment or that they would not be able to afford to buy shares. These motives were followed by two attitudinal variables corresponding to the statements: "I oppose, therefore I will not join" and "I believe that I should do something about climate change, therefore I will join". To a lesser extent, respondents' belief that they would have a say in the project and its management also influenced their willingness to invest in the co-operative scheme.

In response to the question whether they would invest in a wind farm co-operative, 22.5 per cent of respondents said they would not be able to afford to buy shares and 27 per cent neither agreed nor disagreed that they would be able to afford them—hardly negligible proportions considering that a co-operative's aim is to foster participation that will benefit the individuals who join with financial returns and a sense of ownership. The hypothetical price-per-share of £250 indicated in the questionnaire may have been perceived as too high. Nevertheless, this finding strengthens the argument that economic costs must be carefully considered when a co-operative scheme is proposed to a community. Given that no statistical relation was found between the decision to invest and the variables "estimation of income per family member" or "socialcontext2" (a variable introduced in Chapter 5 indicating the relative level of deprivation of each of the surveyed areas) and the decision to invest, the survey results suggest that the cost of shares is considered to be high when compared with alternative purchases, rather than in absolute terms.

Despite the co-operative scheme's possible shortcomings, 52 per cent of respondents considered it an "excellent" or a "good" idea, while just 22 per cent considered it a "bad" or "very bad" idea. Even when respondents were presented with the choice between a co-operative scheme and a community fund scheme, just over 38 per cent preferred the co-operative, 26 per cent the community fund and about 36 per cent stated no preference.

Personal resources

As mentioned in earlier chapters, economic deprivation could be an issue affecting participation in a co-operative scheme. However, multiple deprivation could be an issue for any type of participation—as noted in Chapter 5, just 14 per cent of the returned questionnaires were from the most deprived area surveyed, Bracco, and a significant (at the 0.001 level) though modest correlation ($x = 0.2$) was found between opinion about the wind farm and the level of deprivation.

Income

The picture is mixed regarding income. As mentioned previously, the binomial logistic regression analysis revealed that the affordability of shares was one of the main issues influencing the decision to invest in a wind farm co-operative, yet no statistically significant relation was found between the variables "household income" or "estimation of household income per family member" and the decision to invest. So, if affordability is an issue, it is apparently something that can be perceived as such by households in differing economic situations, who would assess it against competing purchases.

The variable "estimation of income per family member" nevertheless influenced negatively opinion about the wind farm, albeit in a very modest way (Kendall's tau test, $x = -0.111$, $p < 0.014$).

Furthermore, it was found that the variable "estimation of income per family member" was negatively correlated with the variables "benefitscostsvalue" (the scale measuring the subjective perception of costs and benefits related to the proposed wind farm) and "local benefits disadvantages" (respectively, Kendall's tau tests, x = –0.090, p<0.028 and x = –0.144, p<0.003). This suggests that there is a small tendency of respondents in higher income families to see more local and general disadvantages compared with those in lower income families.

Previous research has assumed that more economically affluent areas would lobby more effectively against proposed developments in their locale to protect both the integrity of a perceived highly valuable landscape and their property values (Toke, 2005; van der Horst and Toke, 2010). Analysis of the survey data found only a modest confirmation for the hypothesis that higher income households oppose a proposed wind farm because they wish to protect the integrity of the landscape. In fact, a small significant correlation was found between the variables "estimation of income per family member" and the belief that "the wind farm will look bad on the landscape" (Kendall's tau test, x = –0.124, p<0.008), while no correlation was found between income and the statement about the wind farm causing devaluation of property prices.

Proximity to the wind farm site

Previous studies have indicated that proximity to a proposed wind farm site is a variable that influences acceptability. In this study, merely belonging to either one of the distance bands used to sample the residents was not significantly associated per se with opinions about the wind farm, perhaps because respondents were uncertain as to exactly how far they lived from the site.

While geographical proximity to the site did not influence opinions about the wind farm, the ability to see it from one's home did. A significant, albeit modest, association was found between the variable "seeing the wind farm site from home" and opinion about the proposed wind farm (Cramer's V test, x = 0.272, p<0.000). Similarly, a negative correlation was found between frequency of seeing the site and opinion about the wind farm, but again with a modest coefficient (see Table 5.70, p. 184).

Interestingly, ability to see the wind farm site from home was associated with the variable concerning local benefits and disadvantages (question 10, Appendix A), (Phi test, x = 0.342, p<0.000), suggesting that seeing the site from one's home influenced perceptions of local benefits or disadvantages.

Equally, a negative correlation was found between frequency of seeing the designated site and the perception of local advantages and disadvantages (Kendall's tau test, x = –0.166, p<0.011), implying that respondents who would regularly see the site were likely to predict more disadvantages than benefits from the development.

Other personal resources variables

The questionnaire also comprised other variables that were indicated in previous chapters as personal resources, namely, level of education and knowledge about wind energy. These were not found to be significantly correlated with opinion about the wind farm (a significant correlation between opinion about the wind farm and "knowledge about wind energy" was only found after collapsing the former variable from a five- to a three-answer mode variable).

Nevertheless, a modest correlation can be found between both education level and knowledge about wind energy and opinion about local benefits and dis-advantages. In the case of education (Kendall's tau test, $x = -0.152$, p<0.001), it appeared that respondents with less formal education were slightly more disposed to foresee local advantages than individuals with higher levels of education. Similarly, individuals who were more knowledgeable about wind energy were a little less inclined to anticipate local advantages from the proposed wind farm (Kendall's tau test, $x = -0.094$, p<0.028).

Final remarks

In the conclusions at the end of Chapter 4, it was argued that the qualitative study offered support for the proposed theoretical framework integrating rational choice and attitudes. The same could essentially be said in light of the survey data presented in Chapter 5. Nevertheless, the empirical work dis-cussed in this book indicates that some assumptions about the factors which influence support or opposition to a proposed wind farm development must be revised. One of these, in particular, is that of income: although the research revealed that there is a correlation between income and the perception of costs and benefits associated with a proposed wind farm, it is modest. Moreover, income is not correlated with the specific concern about the decline in property values and is only modestly correlated with concerns about landscape integrity—two specific concerns that both previous literature and the qualitative study presented in this book have suggested are significant to individuals with high levels of income.

"Trust", a contextual factor, appeared as the single primary issue that influ-ences the appraisal of future costs and benefits related to the proposed wind farm, confirming the qualitative study and implying the importance of building this vital feature into local debates and relationships. After trust, ability to see the wind farm from home was the second most important personal resource influencing opinion about costs and benefits associated with the proposed wind farm, a finding that coheres with the previous studies reviewed in Chapter 2 and with the findings of the qualitative survey (see Chapter 4).

The co-operative model was envisioned as a contextual factor that would potentially swing undecided people in favour of the wind farm. After con-sidering the responses obtained by both the qualitative and the quantitative survey, this assumption did not hold. Questionnaire respondents largely answered that a co-operative scheme would not have made a difference in terms of consensus building, thereby lending credibility to those stakeholders that showed scepticism in the qualitative interviews. Yet again, "trust" appeared to be pivotal in shaping opinions of a potential wind farm co-operative, confirming the findings of the interviews.

Finally, it is worth recalling what was said earlier about pro-environmental attitudes. They appear to sustain two antagonistic variants of environmentalism—global and local—which clash in the specific matter of wind farm siting. So, while pro-environmental attitudes might still be considered as influencing behavioural choices in low-cost situations, in the case of acceptability of wind farms during the pre-construction phase, they might lead to both oppositional and supportive attitudes and behaviours.

Policy considerations

The co-operative model cannot in itself overcome local opposition to wind farm proposals, but it can provide opportunities for those who are open-minded about wind energy in their locale, offering them the possibility of gaining revenue from a relatively safe investment. For the community, wind farm co-operatives have the potential to develop a network that will serve it well, strengthening social relations and putting the community in a good place to take advantage of other opportunities.

However, co-operative schemes should not be conceived without some form of community benefit that would give the entire community the ability to derive some advantage as a consequence of hosting a wind farm. Otherwise, this might create a situation in which those who oppose the wind farm, and hence do not join the scheme, will not benefit from the wind farm at all.

As mentioned above, "trust" is a crucial aspect in judgements even about the co-operative model—whether, as in the Westmill case, the main party proposing the co-operative is also the landowner who would benefit from its presence on his land, or somebody else who could be seen to gain some personal advantage from it. This would clearly compromise the possibility of generating trust between the developer (whether this is a co-operative or a commercial entity) and the community. Hence, it makes sense that policymakers should, first, consider making it mandatory for local authorities to assist communities interested in community ownership schemes, as suggested by Haggett and Aitken (2015), by providing information and project management assistance. Second, local authorities should also directly step in, overseeing the process of building community ownership schemes, ensuring that pre-established guidelines are set to provide a fair community space for discussion and deliberation and that these are then followed. This would help prevent, or at least mediate, potential conflicts and help to generate a sense of fairness and ultimately trust in the project.

As several scholars have pointed out (e.g. Haggett and Aitken, 2015; Slee, 2015), there are numerous potential benefits for communities that develop community ownership schemes, but these risk remaining untapped if current national legislation does not change to provide the resources communities need to develop these opportunities. In particular, establishing a mandatory requirement that any wind farm development must provide both a community benefit fund and an opportunity for the local community to co-own at least a share of the project (as current Danish legislation requires; see Slee, 2015) would reduce opposition by ensuring that the perceived costs of wind deployment are minimised, while its perceived potential benefits are maximised.

In this regard, in 2014, the then UK Department of Energy and Climate Change (DECC) outlined a "community energy strategy" (DECC, 2014), which attempted to address the issue of availability of resources to engage in community schemes and aimed to make community ownership widely available for renewable energy developments. A voluntary scheme was encouraged led by the "shared ownership taskforce" (DECC, 2015b), a consultation group including commercial developers and representatives of community organisations. The UK Government stated that if this approach failed, the provision of shared ownership would be made mandatory for any commercial development (DECC, 2015a).

In Scotland, by contrast, the Scottish Government (2018) has recommended, since 2015, that developers offer shared ownership for all renewable energy projects above 50 kW. This position has been further strengthened by a communication by Scotland's chief planner, John McNairney (2015), who, in a letter addressed to the heads of planning at Scottish planning authorities in November 2015, stressed the significance of shared ownership. Although shared ownership does not form part of the material considerations in determining the acceptability of a project in planning terms, it is nevertheless important because the economic and socio-economic impacts of a project are relevant material considerations.

In conclusion, while a growing number of policymakers now understand the importance of community benefits and community ownership, a number of ambiguities remain regarding both the policy discourse and national legislation concerning these schemes. A clear policy choice that makes community benefits and the provision of shared ownership schemes mandatory for each renewable energy project would help in making renewable developments more acceptable, and in enabling a positive role for citizens in the energy transition.

National legislation making community and shared ownership schemes mandatory would also support an energy justice policy approach based on the tenets of distributional and procedural justice (McCauley *et al.*, 2013), thereby advancing an egalitarian view of energy policy (Pellegrini-Masini, 2019; Pellegrini-Masini *et al.*, 2020).

References

Batel, S. and Devine-Wright, P. 2015. Towards a better understanding of people's responses to renewable energy technologies: Insights from social representations theory. *Public Understanding of Science*, 24, 311–325.

Bauwens, T., Gotchev, B. and Holstenkamp, L. 2016. What drives the development of community energy in Europe? The case of wind power cooperatives. *Energy Research and Social Science*, 13, 136–147.

Bell, D., Gray, T., Haggett, C. and Swaffield, J. 2013. Re-visiting the "social gap": Public opinion and relations of power in the local politics of wind energy. *Environmental Politics*, 22, 115–135.

De Groot, J. I. and Steg, L. 2009. Mean or green: Which values can promote stable pro-environmental behavior? *Conservation Letters*, 2, 61–66.

DECC. 2011. *Renewable energy roadmap*. Department of Energy and Climate Change, London. Available at: www.gov.uk/government/publications/renewable-energy-road-map.

DECC. 2014. *Community energy strategy*. Department of Energy and Climate Change, London. Available at: www.gov.uk/government/uploads/system/uploads/attachment_data/file/275163/20140126Community_Energy_Strategy.pdf.

DECC. 2015a. *Community energy strategy: Update*. Department of Energy and Climate Change, London. Available at: www.gov.uk/government/uploads/system/uploads/attachment_data/file/414446/CESU_FINAL.pdf.

DECC. 2015b. *Government response to the Shared Ownership Taskforce*. Department of Energy and Climate Change, London. Available at: www.gov.uk/government/publications/government-response-to-the-shared-ownership-taskforce.

Diekmann, A. and Preisendörfer, P. 2003. Green and greenback: The behavioral effects of environmental attitudes in low-cost and high-cost situations. *Rationality and Society*, 15, 441–472.

European Parliament. 2018. Directive (EU) 2018/2001 of the European Parliament and of the Council of 11 December 2018 on the promotion of the use of energy from renewable sources. *Official Journal of the European Union*, 128, 83–206.

Haggett, C. and Aitken, M. 2015. Grassroots energy innovations: The role of community ownership and investment. *Current Sustainable/Renewable Energy Reports*, 2, 98–104.

Kalkbrenner, B. J. and Roosen, J. 2016. Citizens' willingness to participate in local renewable energy projects: The role of community and trust in Germany. *Energy Research & Social Science*, 13, 60–70.

Lipp, J. and McMurtry, J. J. 2015. *Benefits of renewable energy co-operatives: Summary of literature review from the Measuring the Co-operative Difference Research Network*. Measuring the Co-operative Difference Research Network, Canada. Available at: http://ec.msvu.ca:8080/xmlui/handle/10587/1608.

Maslow, A. H. 1987. *Motivation and personality*. New York: Harper & Row.

McCauley, D., Heffron, R. J., Stephan, H., Jenkins, K., Gillard, R., Snell, C. and Bevan, M. 2013. Advancing energy justice: The triumvirate of tenets and systems thinking. *International Energy Law Review*, 32, 107–110.

McNairney, J. 2015. Energy targets and Scottish planning policy. Local Government and Communities Directorate, Scottish Government, Edinburgh. Available at: www.gov.scot/publications/energy-targets-and-scottish-planning-policy-chief-planner-letter/.

Misztal, B. 2013. *Trust in modern societies: The search for the bases of social order*. Oxford: John Wiley & Sons.

Pellegrini-Masini, G. 2007. The carbon-saving behaviour of residential households. Futures of Cities—51st International Federation of Housing and Planning World Congress, 23–26 September 2007, Copenhagen.

Pellegrini-Masini, G. 2019. Energy equality and energy sufficiency: New policy principles to accelerate the energy transition. European Council for an Energy Efficient Economy 2019 Summer Study, "Energy efficiency first, but what next?", 3–8 June, Belambra Presqu'ile de Glens, France, 143–148.

Pellegrini-Masini, G., Pirni, A. and Maran, S. 2020. Energy justice revisited: A critical review on the philosophical and political origins of equality. *Energy Research & Social Science*, 59, 101310. Available at: https://doi.org/10.1016/j.erss.2019.101310.

Perlaviciute, G. and Steg, L. 2014. Contextual and psychological factors shaping evaluations and acceptability of energy alternatives: Integrated review and research agenda. *Renewable and Sustainable Energy Reviews*, 35, 361–381.

Scottish Government. 2018. *Good practice principles for shared ownership of renewable energy developments*. Energy and Climate Change Directorate, Edinburgh. Available at: www.gov.scot/binaries/content/documents/govscot/publications/consultation-paper/2018/11/consultation-scottish-government-good-practice-principles-shared-ownership-renewable-energy-developments/documents/scottish-government-good-practice-principles-shared-ownership-renewable-energy-developments/scottish-government-good-practice-principles-shared-ownership-renewable-energy-developments/govscot%3Adocument/00543554.pdf.

Slee, B. 2015. Is there a case for community-based equity participation in Scottish on-shore wind energy production? Gaps in evidence and research needs. *Renewable and Sustainable Energy Reviews*, 41, 540–549.

Steg, L., Bolderdijk, J. W., Keizer, K. and Perlaviciute, G. 2014. An integrated framework for encouraging pro-environmental behaviour: The role of values, situational factors and goals. *Journal of Environmental Psychology*, 38, 104–115.

Taormina, R. J. and Gao, J. H. 2013. Maslow and the motivation hierarchy: Measuring satisfaction of the needs. *American Journal of Psychology*, 126, 155–177.

Toke, D. 2005. Explaining wind power planning outcomes: Some findings from a study in England and Wales. *Energy Policy*, 33, 1527–1539.

van der Horst, D. and Toke, D. 2010. Exploring the landscape of wind farm developments: Local area characteristics and planning process outcomes in rural England. *Land Use Policy*, 27, 214–221.

van Veelen, B. and Haggett, C. 2016. Uncommon ground: The role of different place attachments in explaining community renewable energy projects. *Sociologia Ruralis*, 57, 533–554.

Walker, B. J. A., Wiersma, B. and Bailey, E. 2014. Community benefits, framing and the social acceptance of offshore wind farms: An experimental study in England. *Energy Research and Social Science*, 3, 46–54.

Warren, C., Lumsden, C., O'Dowd, S. and Birnie, R. 2005. "Green on green": Public perceptions of wind power in Scotland and Ireland. *Journal of Environmental Planning and Management*, 48, 853–875.

Conclusions

This research project set out to answer the following research questions:

1 Which factors influence the acceptability of wind farms? How do they relate to one another?
2 Which factors influence participation in wind farm co-operatives? How are they related to one another?
3 Is the co-operative model effective in eliciting the participation of local communities and in overcoming opposition towards wind developments? Why?
4 Do individuals perceive their status as citizens as a source of moral obligation to protect the environment? (In other words, is environmental citizenship perceived as a source of moral obligation to protect the environment?)

In order to answer these research questions, a theoretical and empirical contribution was developed that elaborated on and collected evidence about what influences the acceptability of wind farms and how the co-operative model can play a role in this regard. The choice to conceptualise an interpretative theoretical framework that integrates rational choice theory, attitudinal theories and resource-based models fitted with the data collected reasonably well, and demonstrates the potential for its further employment when researching other environmentally significant behaviours.

With regard to the first research question, it was shown that a proposed wind farm's perceived costs and benefits influence its acceptability. In particular, these costs and benefits (in decreasing order of importance) were: the fear that the wind farm would harm the health of the local community; the belief that the wind farm would help with climate change; the belief that the wind farm would be noisy; the opinion that the wind farm would create an eyesore on the landscape; and, finally, the belief that the wind farm would alleviate fuel dependency. Residents' personal and social resources, including "trust in the developers", "seeing the wind farm site from home" and "information about the wind farm" further influenced how they perceived the costs and benefits of a proposed wind farm. Trust, in particular, stood out as a critical variable that influenced the attitudes of respondents towards the locally proposed wind farm and the perception of its related costs and benefits.

Attitudes towards the environment and the place (i.e. place attachment) also played a role but were minor factors influencing residents' opinions about a wind farm. In particular, individuals who displayed pro-environmental attitudes were not uniformly concerned about protection of the global environment and, hence, in support of a wind farm. Rather, a considerable number demonstrated greater concern about the integrity of their local environment.

In answering the second and third research questions, it was found that a co-operative model in and of itself was not sufficient to convince undecided residents to support a locally proposed wind farm. Many regarded the co-operative as potentially divisive; nevertheless, respondents expressed predominantly favourable views towards the scheme and preferred it over a community benefits scheme.

The issues that appeared to influence opinions about the co-operative model the most, in decreasing order of importance, were: "trust", i.e. the fear that the scheme might be a "ploy" to facilitate support for the wind farm; the expectation that the co-operative's revenue would benefit the community; the expectation that the co-operative would create social capital; and, finally, the concern that the co-operative would offer the worst compensation to its opponents.

The main reasons, from greatest to least importance, that influenced the intention to invest (or not) appeared to be: the consideration of the investment as a good opportunity; the respondent's belief that they would not be able to afford shares; the fact of opposing the wind farm; the belief that something must be done about climate change; and, finally, the belief that investment would give the respondent a say in the co-operative project and its management.

With regard to "environmental citizenship" (the fourth research question), while respondents largely felt a sense of responsibility towards the environment, they were polarised between preference for protection of the global environment and protection of the local environment, and disregarded the national environmental dimension or the option of rejecting any responsibility for environmental matters. Nevertheless, this variable did not appear to be significantly correlated with opinion about the wind farm.

In Chapter 6 several policy considerations concerning the co-operative model were presented that are worth briefly reiterating here. While it is true that the co-operative model of local ownership of wind farms is not a guaranteed means of winning over opponents—and possibly even those who are undecided—it was nonetheless viewed favourably by many in this study. Together with a community benefits scheme (which can be complementary), a co-operative scheme might have a place among a number of community ownership options that could be proposed to a local community to provide increased opportunities to benefit from renewable energy. The co-operative model has the potential to provide a sense of ownership, strengthen community networks, raise community awareness of energy issues and to better position communities to address future collective local issues or opportunities.

All things considered, the fact that the Scottish and UK governments, among others, have recommended that all developers offer local communities the possibility of shared ownership must be considered a positive development. Nevertheless, it could be argued that local shared ownership should be a mandatory requirement for any wind farm planning application, and that precise guidelines on its implementation should be set and followed by all developers, under the supervision of the competent local authorities. Such legislation would be likely to be effective in establishing the trust between the developer and the community that is key to avoiding divisive local controversies.

This research and its findings can also be considered in the wider context of the energy transition. While the emphasis of the transition is often placed on the technological transformation of energy systems and decarbonisation, it is becoming clearer to policymakers that this needs to be equitable (European Union, 2019; World Energy Council, 2019). The already vast literature on energy justice has argued, in the last ten years, that just energy policies need to rest on distributional justice and procedural justice (Guruswamy, 2010; McCauley *et al.*, 2013; Sovacool *et al.*, 2016), which are both ultimately based on equalitarian considerations (Pellegrini-Masini *et al.*, 2020). But translating these principles into policy means challenging the established institutional contexts of the current socio-technical system whose structured dynamics appear difficult to change or, even, "locked-in" (Geels, 2005). Change, then, can happen both through bottom-up social innovations that develop in niches and top-down steering by governments (Grin *et al.*, 2010).

Whether governments try to support gradual change in the energy sector by stimulating niche energy social innovations or pursuing more rapid and abrupt top-down driven change, they will have to face issues of participation and acceptability by the public and stakeholders. If the research that I have presented in this book can lend itself to some reflections on wider policy implications, these could be summed up in the following considerations.

As noted earlier (Pellegrini-Masini, 2007; Pellegrini-Masini and Leishman, 2011; Perlaviciute and Steg, 2014; Steg *et al.*, 2014) and confirmed by the findings of this research, the perception of benefits and costs, including the benefit of consistency with one's environmental (and other) attitudes, will have a profound impact on the behavioural and organisational choices of actors to accept, reject or participate in energy social innovations or policies. This will alter established patterns of energy consumption and production. Energy policies will need to change the perception of benefits and costs of alternative energy choices, and possibly strengthen pro-environmental attitudes, through a complementary approach that aims to strengthen extrinsic and intrinsic motivations to act pro-environmentally (Frey, 1999).

Trust is a pivotal element influencing energy social innovations (see, e.g., Pellegrini-Masini *et al.*, 2019) and community energy schemes (Walker *et al.*, 2010) in general. This research confirmed that trust is a crucial variable that affects local opinion about proposed wind farms and co-operative schemes. It appears to be linked to perceptions of fairness and procedural justice (Jagers *et al.*, 2010; Huijts *et al.*, 2012, Drews and van den Bergh, 2016,). Therefore, policies encouraging the energy transition should aim to be fair and inclusive, based on procedural (formal) equality, not only because this could be argued to be morally desirable (Pellegrini-Masini *et al.*, 2020) but because they appear to motivate citizens towards acceptance and participation (Drews and van den Bergh, 2016).

Finally, holding together procedural justice and fairness while shaping energy policy related benefits and costs in order to build extrinsic motivations (along with intrinsic ones) leads to the choice of implementing distributionally just policies which would favour the many over the few. Energy equality along with energy sufficiency (Pellegrini-Masini, 2019) will need to drive the transition if the current energy injustices, often related to economic inequality at the global level (Gore, 2015) and within countries (Galvin and Sunikka-Blank, 2018; Galvin, 2019), are to be addressed. Distributional justice policies leading towards more egalitarian societies will most likely support a per capita decrease in carbon emissions in high-income countries (Jorgenson *et al.*, 2016; Knight *et al.*, 2017), and might increase pro-environmental behaviours (Carlisle and Smith, 2005; Wilkinson and Pickett, 2010), while increasing quality of life in general (Wilkinson *et al.*, 2010). In this respect, citizens' support for community benefit schemes and the co-operative model of shared ownership of wind farms confirmed in this research supports the idea that distributionally just energy policies at the community level (and beyond) are both desirable and necessary.

References

Carlisle, J. and Smith, E. R. 2005. Postmaterialism vs. egalitarianism as predictors of energy-related attitudes. *Environmental Politics*, 14, 527–540.

Drews, S. and van den Bergh, J. C. J. M. 2016. What explains public support for climate policies? A review of empirical and experimental studies. *Climate Policy*, 16, 855–876.

European Union. 2019. *Clean energy for all Europeans*. Directorate-General for Energy, Luxembourg. Available at: https://publications.europa.eu/en/publication-detail/-/publication/b4e46873-7528-11e9-9f05-01aa75ed71a1/language-en.

Frey, B. S. 1999. Morality and rationality in environmental policy. *Journal of Consumer Policy*, 22, 395–417.

Galvin, R. 2019. Letting the Gini out of the fuel poverty bottle? Correlating cold homes and income inequality in European Union countries. *Energy Research and Social Science*, 58, 101255. Available at: https://doi.org/10.1016/j.erss.2019.101255.

Galvin, R. and Sunikka-Blank, M. 2018. Economic inequality and household energy consumption in high-income countries: A challenge for social science based energy research. *Ecological Economics*, 153, 78–88.

Geels, F. W. 2005. Processes and patterns in transitions and system innovations: Refining the co-evolutionary multi-level perspective. *Technological Forecasting & Social Change*, 72, 681–696.

Gore, T. 2015. *Extreme carbon inequality: Why the Paris climate deal must put the poorest, lowest emitting and most vulnerable people first*. Oxfam International, Oxford. Available at: www.oxfam.org/sites/www.oxfam.org/files/file_attachments/mb-extreme-carbon-inequality-021215-en.pdf.

Grin, J., Rotmans, J. and Schot, J. 2010. *Transitions to sustainable development: New directions in the study of long term transformative change*. Abingdon, UK: Routledge.

Guruswamy, L. 2010. Energy justice and sustainable development. *Colorado Journal of International Environmental Law and Policy*, 21, 231–275.

Huijts, N. M. A., Molin, E. J. E. and Steg, L. 2012. Psychological factors influencing sustainable energy technology acceptance: A review-based comprehensive framework. *Renewable and Sustainable Energy Reviews*, 16, 525–531.

Jagers, S. C., Löfgren, Å. and Stripple, J. 2010. Attitudes to personal carbon allowances: Political trust, fairness and ideology. *Climate Policy*, 10, 410–431.

Jorgenson, A. K., Schor, J. B., Knight, K. W. and Huang, X. 2016. Domestic inequality and carbon emissions in comparative perspective. *Sociological Forum*, 31, 770–786.

Knight, K. W., Schor, J. B. and Jorgenson, A. K. 2017. Wealth inequality and carbon emissions in high-income countries. *Social Currents*, 4, 403–412.

McCauley, D., Heffron, R. J., Stephan, H., Jenkins, K., Gillard, R., Snell, C. and Bevan, M. 2013. Advancing energy justice: The triumvirate of tenets and systems thinking. *International Energy Law Review*, 32, 107–110.

Pellegrini-Masini, G. 2007. The carbon-saving behaviour of residential households Futures of Cities—51st International Federation of Housing and Planning World Congress, 23–26 September 2007, Copenhagen.

Pellegrini-Masini, G. 2019. Energy equality and energy sufficiency: New policy principles to accelerate the energy transition. European Council for an Energy Efficient Economy 2019 Summer Study, "Energy efficiency first, but what next?", 3–8 June, Belambra Presqu'ile de Glens, France, 143–148.

Pellegrini-Masini, G. and Leishman, C. 2011. The role of corporate reputation and employees' values in the uptake of energy efficiency in office buildings. *Energy Policy*, 39, 5409–5419.

Pellegrini-Masini, G., Macsinga, I., Albulescu, P., Löfström, E., Sulea, C., Dumitru, A. and Nayum, A. 2019. *D6.1 Report on social innovation drivers, barriers, actors and network structures* (H2020 Project SMARTEES Grant Agreement No. 763912). Trondheim, Norway.

Pellegrini-Masini, G., Pirni, A. and Maran, S. 2020. Energy justice revisited: A critical review on the philosophical and political origins of equality. *Energy Research & Social Science*, 59, 101310. Available at: https://doi.org/10.1016/j.erss.2019.101310.

Perlaviciute, G. and Steg, L. 2014. Contextual and psychological factors shaping evaluations and acceptability of energy alternatives: Integrated review and research agenda. *Renewable and Sustainable Energy Reviews*, 35, 361–381.

Sovacool, B. K., Heffron, R. J., McCauley, D. and Goldthau, A. 2016. Energy decisions reframed as justice and ethical concerns. *Nature Energy*, 1, 16024.

Steg, L., Bolderdijk, J. W., Keizer, K. and Perlaviciute, G. 2014. An integrated framework for encouraging pro-environmental behaviour: The role of values, situational factors and goals. *Journal of Environmental Psychology*, 38, 104–115.

Walker, G., Devine-Wright, P., Hunter, S., High, H. and Evans, B. 2010. Trust and community: Exploring the meanings, contexts and dynamics of community renewable energy. *Energy Policy*, 38, 2655–2663.

Wilkinson, R. G. and Pickett, K. 2010. *The spirit level: Why greater equality makes societies stronger*. New York: Bloomsbury.

Wilkinson, R., Pickett, K. and De Vogli, R. 2010. Equality, sustainability, and quality of life. *The BMJ*, 341, 1138–1140.

World Energy Council. 2019. *World energy trilemma index report 2019*. World Energy Council, London. Available at: www.worldenergy.org/assets/downloads/WETrilemma_2019_Full_Report_v4_pages.pdf.

Appendix A

The postal survey questionnaire

1 What is the highest educational level that you have attained?

TABLE A

❑ primary school	❑ graduate level	❑ professional/vocational
❑ secondary school	❑ postgraduate level	

2 How many people are there in your household including yourself? Please write the number …

3 What is your approximate household income before tax each year?

TABLE B

❑ under £10,000 per year	❑ £20,000–£29,999	❑ £50,000–£79,999
❑ £10,000–£19,999	❑ £30,000–£49,999	❑ £80,000 or more

4 Please say what you think are the correct answers to the following questions about wind power.

4.1 How much pollution do wind turbines produce in comparison with coal-fired power stations?
   ❑ more pollution   ❑ less pollution   ❑ I don't know

4.2 Is the electricity produced by wind turbines cheaper or more expensive to produce than electricity produced by other means such as coal-fired power stations?
   ❑ cheaper   ❑ more expensive   ❑ I don't know

4.3 Whatever the location, do wind turbines produce a steady stream of electricity?
   ❑ yes   ❑ no   ❑ I don't know

4.4 When people talk about renewable energy, do they consider wind power to be a type of renewable energy?
   ❑ yes   ❑ no   ❑ I don't know

5 Are you aware that the ["…"] wind farm has been proposed in your local area?
   ❑ yes   ❑ no

6 Can you see from your home ["…"], the site of your proposed local wind farm?
   ❑ yes   ❑ no   ❑ I don't know where the site is
   6.1 If you answered NO: How often do you see the wind farm proposed site?
        ❑ never   ❑ rarely   ❑ sometimes   ❑ often   ❑ very often
7 What do you think of the presence of this wind farm in your area?
   I … strongly disagree ❑ disagree   ❑ neither agree nor disagree
   ❑ agree   ❑ strongly agree ❑
   7.1 Why? Please briefly explain the reasons for your choice (please write)

   ....
   ....
   ....
   ....
   ....
   ....

   Please say if you AGREE or DISAGREE with the following statements:

8 I trust the developers of the wind farm in the way they deal and have dealt
   with the local community: strongly disagree ❑ disagree ❑ neither agree
   nor disagree ❑ agree ❑ strongly agree ❑
9 I feel that I have been thoroughly informed about the wind farm: strongly
   disagree ❑ disagree ❑ neither agree nor disagree ❑ agree ❑ strongly
   agree ❑
10 Please tick the box of the ONE statement that you agree with:
   I think that the local wind farm will bring LOCALLY …
   many benefits and no disadvantages   ❑
   more benefits than disadvantages   ❑
   neither benefits nor disadvantages   ❑
   more disadvantages than benefits   ❑
   many disadvantages and no benefits   ❑

   Please read the following statements and say if you AGREE or DISAGREE

Please read the following information before continuing to fill in the questionnaire.

• A **community wind farm co-operative** is a form of community ownership
   of wind farms. In this scheme the revenue from the wind farm goes to individual shareholders, the members of the co-operative, who are local residents that bought one or more shares. A minor part of the revenue can be
   spent on community projects. A co-operative is owned by its members and
   every member has the same right to have a say in the business regardless of
   the number of shares that he or she owns.

TABLE C

The wind farm will ...	Strongly disagree	Disagree	Neither agree nor disagree	Agree	Strongly agree
11 ... harm the health of my community					
11.1 ... help to fight climate change					
11.2 ... look bad on the landscape					
11.3 ... improve the local economy					
11.4 ... bring down the local property prices					
11.5 ... attract tourists					
11.6 ... be unpleasantly noisy					
11.7 ... generate costlier electricity than if it was generated by ordinary fuels					
11.8 ... help to free the country from dependence on foreign fuels					

- A **community fund** is another option to benefit the local community. It is not a form of ownership but is set up by the wind farm developer, who agrees to pay into the fund a sum of money which the community will spend on projects for its own benefit. It will be managed on behalf of the community by a community trust or by the local authority.

12 Are you aware of any of the following schemes being proposed near you?
a 'community fund' sponsored by the developer of the wind farm ❑
a 'community wind farm co-operative' to allow residents to buy shares in the wind farm ❑
a different scheme from those listed, which is (please specify) ....

....

....

I am not aware of any scheme ❑

If a 'community wind farm co-operative' was proposed for your area, would you AGREE or DISAGREE with the following statements?

TABLE D

The co-operative ...	Strongly disagree	Disagree	Neither agree nor disagree	Agree	Strongly agree
13. ... would just be a ploy to buy residents' consensus					
13.1 ... would give locals the chance to benefit from the revenue of the wind farm					
13.2 ... would create a permanent divide in the local community between those who would join and those who would oppose the wind farm					
13.3 ... would persuade those who are undecided to support the wind farm					
13.4 ... would offer the worst compensation for those who oppose the wind farm because their decision not to join means they would not receive any revenue					
13.5 ... would involve local people not only financially but also in its management: it would create a stable network of local residents who might support further community activities and projects					
13.6 ... would persuade even opponents of the wind farm to accept the development					
13.7 ... wouldn't make any difference. People would support or oppose the wind farm regardless of whether there is a co-operative scheme or not					

14 Would you invest in a 'community wind farm co-operative', if this was proposed for your local wind farm and the minimum requested investment was £250?
yes ❑ no ❑

15 Please give reasons for your last answer: look at the following list of statements and say if you AGREE or DISAGREE

TABLE E

	Strongly disagree	Disagree	Neither agree nor disagree	Agree	Strongly agree
15.1 I think that it would be a good investment opportunity					
15.2 I oppose the wind farm so I would never join in					
15.3 I believe that we all should do something to fight climate change, therefore I would join					
15.4 I couldn't afford to buy the shares					
15.5 If people around me, in my community, would support it, so would I					
15.6 I don't care about the wind farm and so I would not care about the co-operative					
15.7 I would be able to have a say in the development of the wind farm and its management					

15.8 None of the above, the reason being (please write).........................

....

....

16 All in all, what do you think of a 'community wind farm co-operative'?

TABLE F

❏ it is a very bad idea	❏ I have no opinion on this	❏ it is an excellent idea
❏ it is a bad idea		❏ it is a good idea

17 If you could choose ONE of the schemes that we have presented for your local wind farm, which one would you choose? Please mark your answer of choice, please choose only ONE answer.

- a 'community wind farm co-operative' in which residents invest to buy affordable shares and therefore receive an annual revenue ❏
- a 'community fund' sponsored by the developer/owner of the wind farm which will fund community collective projects (e.g. a community hall) ❏
- I have no preference ❏

Which ONE of the following statements describes best what you think?
I think that …

18 We all have to do something to protect the global environment because we all share planet earth ❏
   18.1 We all have to do something to take care of our country's environment because that is in our best interest ❏
   18.2 We all have to do something to protect our local environment, because it's the place where we live with our families, in our communities ❏
   18.3 It's a matter for the government, not us, to take care of the environment ❏

Please tick the box of the ONE statement that you agree the most with:

19 The environment should be the priority of the government even if this means damaging the economy ❏
   19.1 The environment should be the priority of the government but this should not damage the economy ❏
   19.2 The economy should be the priority of the government but this should not damage the environment ❏
   19.3 The economy should be the priority of the government even if this means damaging the environment ❏

Please say if you AGREE or DISAGREE with the following statements

TABLE G

	Strongly disagree	Disagree	Neither agree nor disagree	Agree	Strongly agree
20. I like how my area looks					
20.1 I like my community					

Appendix B

The qualitative study interview guide

Interview guide

The following introduction was used:

> I am a PhD student. My research regards acceptability of wind farms at the local community level. There is currently a debate in the country about a possible larger deployment of renewable energy, particularly regarding onshore wind farms. The great importance of this debate is evident for the energy future of Britain. I am going to ask you some questions related with this now. I wish to underline that there are no right or wrong answers to my questions, so please feel completely free to express any opinion.
>
> All your answers will be treated as confidential and you can decide to withdraw at any stage of the interview. I would find it useful to record the interview in order to report exactly your opinions. Do you agree?

Warm-up question

This question is posed to allow the interviewee to start the interview comfortably. The topic is trivial, but it will also be useful for the interviewer to ascertain the respondent's perception of himself/herself and his/her role with regards to the wind farm.

> Could you please describe yourself and your role in your group/organisation/institution?

I – Responsibility, citizenship, participation

These questions refer to research question 4: "Do individuals perceive their status as citizens as a source of moral obligation to protect the environment? (In other words, is environmental citizenship perceived as a source of moral obligation to protect the environment?)".

1 Who or what do you think is responsible for causing climate change?
  Probe: Is that your personal view or your organisation's?

This probe will be repeated throughout the questionnaire whenever it is needed to ascertain if the opinion presented is a personal position or if it represents the views of the group/institution to which the respondent belongs. In case this is not made explicit by the interviewee, it was assumed that s/he was presenting her/his personal views.

2 Have you ever come across the concept of citizenship? What does it mean to you?

If needed, the concept of citizenship was essentially recalled with reference to the environment to allow the respondent to see the connection between the citizenship and the environmental debate.

> Citizenship is the membership of the national political community. Citizens as a consequence hold rights and duties towards the community. In this respect, environmental quality is a public good because it doesn't completely belong to single users, e.g. just think about the air quality, we all consume the same air. Therefore, as for other public goods (e.g. public health, security) we, as citizens, have environmental rights and duties—e.g. the right to the availability of clean air and the duty to avoid illegal dumping.

3 What is your opinion about citizens' rights and duties in relation to the environment?
Prompts: moral rights and duties—legal rights and duties
Probe: How effective are they to protect the environment?
4 If the environment belongs to everyone, who should take care of it?
Prompts: government—businesses—British citizens
5 What do you think about people being producers of renewable energy as a way to protect the environment?
6 What do you think of local community involvement in protecting the environment? Why?
Prompts:
a knowledge of the local environment
b local interests
7 Could you give an example of community involvement?
Prompt: Westmill?

II – Community involvement, social-enterprises/ co-operatives and wind farms

The following questions aim to answer the research question: "Which attitudes are held by stakeholders and citizens towards the social enterprise and community owned schemes of wind farms?".[1]

Westmill will be a social enterprise, a co-operative in particular.

8   What do you think of social enterprises?
9   And what do you think about co-operatives?

If the respondent asked for a definition of "social enterprise", the following was provided by the Department of Trade and Industry, which defines a social enterprise as *"... a business with primarily social objectives whose surpluses are principally reinvested for that purpose in the business or in the community, rather than being driven by the need to maximise profit for shareholders and owners".*
If the respondent didn't have a clear understanding of the meaning of "co-operative" the following definition by the Department of Trade and Industry was provided. The DTI defines co-operatives as *"... independent, democratically controlled enterprises. They are owned and governed by their members, with the aim of meeting common social, economic and environmental needs".*

10   Are there any co-operatives in your local area? Can you tell me something about them?
11   How are cooperatives seen by members of the public?
       Prompts: positive/good; negative/bad
12   What do you think are the advantages or disadvantages of community owned co-operatives that produce green electricity, e.g. a wind farm?
       Prompts:
       a   Are you aware of the Danish experience? Do you think it could be repeated in the UK? Why? Why not?
       b   Are you aware of Baywind?
13   What do you think of community owned co-operatives of renewable energy as a means to tackle climate change?
14   Do you think that renewable energy developments owned by the local community have or don't have the same local impact as those owned by non-local companies? Why?
       Probes:
       What about the ...
       a   ... local economic impact?
       b   ... social impact on community life?
       c   ... environmental impact?
15   Have you ever considered the relative benefits of cheaper energy versus locally owned energy. Which of these is more desirable? Why?

III – Wind farm opposition and community owned co-operative schemes

The questions in this section aimed to answer research question 3: "Is the cooperative scheme effective in eliciting the participation of local communities and in overcoming opposition towards wind developments? Why?".

16 Have you ever thought about any possible link between opposition or support and local ownership of a wind farm?

If not clear to the interviewee, it was rephrased as follows:

Do you think that local ownership of wind farms may influence levels of opposition or support for wind farms? How?

17 Wind turbines are sometimes controversial and local opposition is often present. Could the local ownership of the wind farm with its revenue put back into the local community compensate for the perception of local negative consequences of a wind farm, such as noise and visual impact?

IV– Perception of factors influencing participation in community owned co-operatives

This question aimed to answer research question 2: "Which factors influence participation in farm cooperatives? How are they related to one another?".

18 Have you ever discussed in your group/organisation/institution what could motivate or prevent local residents to purchase shares in a local co-operative wind farm? What do you think about it?

It was impossible to prompt all the factors raised in the literature review; therefore, leaving the question completely open appeared to be the most reasonable choice.

Note

1 During the development of the research project it was decided not to pursue this research question further.

Index

Page numbers in **bold** denote tables, those in *italics* denote figures.